GMP Inspections

Basics for Beginners

ISBN-13: 978-1548715328

ISBN-10: 1548715328

www.validationresources.org

Contents

CHAPTER 1

Good Manufacturing Practices

CHAPTER 2

Preparation for Audits

CHAPTER 3

Inspection of Quality Systems

CHAPTER 4

During the Inspection

CHAPTER 5

Biotechnology Inspection Guide

CHAPTER 6

Medical Device Inspection Guide

CHAPTER 7

Sterile Drugs Inspection Guide

CHAPTER 8

Computerised Systems Inspection Guide

CHAPTER 9

Guide to Inspection Of
Validated Cleaning Processes

"Coming together is a beginning. Keeping together is progress. Working together is success."

-Henry Ford -

CHAPTER 1

Good Manufacturing Practices (GMP)

Introduction

Good Manufacturing Practices are a set of practices that are required in order to comply with industry standards and regulations. GMP helps to minimise the risks involved during manufacturing and helps to ensure products meet quality and regulatory standards. A GMP quality system ensures that products are consistently produced and controlled according to predefined quality standards. It is designed to minimise the risks involved in any pharmaceutical production that cannot be eliminated through testing the final product.

CGxP /cGMP

Often, a broader term is used in industry – GxP - where the "x" is used as an umbrella letter representing different subjects or disciplines in industry. Some prime examples include GLP (Good Laboratory Practice), GDP (Good Documentation Practice), GEP (Good Engineering Practice) and GMP (Good Manufacturing Practices). Furthermore, the use of a lowercase "c" as a prefix indicates "current" or "up-to-date". So cGMP stands for Current Good Manufacturing Practices.

This means that some conventions or practices are subject to change within the industry. Therefore, it is important to be up to date in the application of cGxP or cGMP

There are multiple regulators and organisations that provide definitions of "Good Manufacturing Practices". They include organisations such as the World Health Organisation (WHO) and the International Society of Pharmaceutical Engineering (ISPE), PIC/s, EU Eurdralex Volume 4, Good Manufacturing Pracatices. Other definitions are offered by bodies such as the American competent authority for Food and Drug Administration. It is good to have an awareness of how organisations, bodies and competent authorities define GMP, and one should always review the local regulatory landscape. Below some definitions are provided to provide a feel for GMP and highlight the common thread between definitions.

W.H.O. World Health Organisation: "Good Manufacturing Practices (GMP, also referred to as 'cGMP' or 'current Good Manufacturing Practice') is the aspect of quality assurance that ensures that medicinal products are consistently produced and controlled to the quality standards appropriate to their intended use and as required by the product specification."

Food and Drug Administration: cGMP refers to the Current Good Manufacturing Practice regulations enforced by the US Food and Drug Administration (FDA). cGMPs ensure systems are properly designed and monitored, safeguarding the control of manufacturing processes and facilities. Adherence to the cGMP regulations ensures the identity, strength, quality and purity of drug products by requiring that manufacturers of medications adequately control manufacturing operations. This includes establishing strong Quality Management Systems, obtaining appropriate quality raw materials, establishing robust operating procedures, detecting and investigating product quality deviations and maintaining reliable testing laboratories. This formal system of controls at a pharmaceutical company, if adequately put into practice, helps to prevent instances of contamination, mix-ups, deviations, failures and errors. This assures that drug products meet their quality standards.

MHRA (Medicines and Healthcare Products Regulatory Agency) defines GMP as follows:

"Good Manufacturing Practice (GMP) is that part of quality assurance which ensures that medicinal products are consistently produced and controlled to the quality standards appropriate to their intended use and as required by the marketing authorisation (MA) or product specification. GMP is concerned with both production and quality control. Many of the drivers of GMP in effect are also benefits to the manufacturer. Good manufacturing practices are an expected practice in regulated industries and a manufacturer must meet all relevant GMP regulations if they wish to manufacture within a country or sell to a particular market. It is important to maintain accurate, complete, up-to-date and consistent information to ensure patient safety and reduce any potential risks."

A basic tenet of GMP is that (1) quality cannot be tested into a batch of product and (2) quality must be built into each batch of product during all stages of the manufacturing process.

Good Manufacturing Practice (GMP) describes a set of principles and procedures that when followed helps ensure that therapeutic goods are of high quality.

There are different codes of GMP, depending on the type of therapeutic good:

- ➢ Good Manufacturing Practice for Medicines
- ➢ Good Manufacturing Practice for Human Blood and Tissues
- ➢ A different system, known as conformity Assessment, is used to ensure that medical devices are of high quality.

The following section shows the structure and key headings as they appear in the respective sources. GMP requirements are based on both regulatory authorities and other international organisations such as PICS/s, WHO, FDA etc.

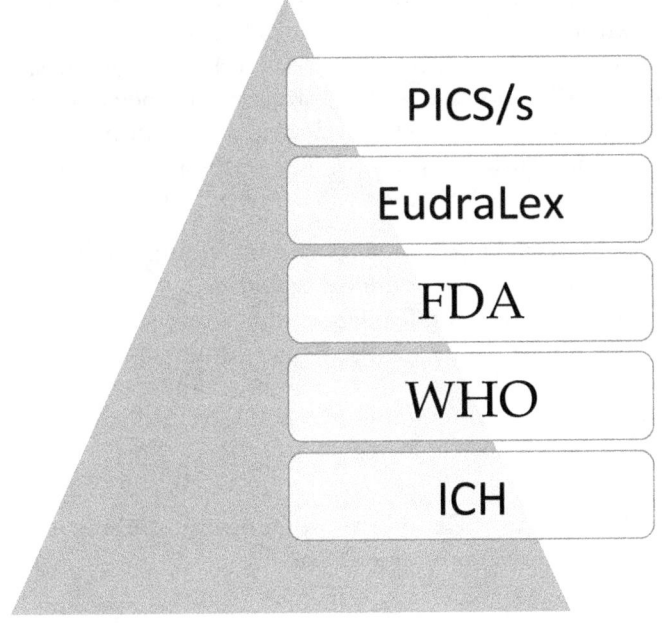

Figure: Organisations and bodies that publish GMP requirements

PICS/s Manufacturing Principles for Medicinal Products:

Pharmaceutical Inspection Convention and Pharmaceutical Inspection Co-Operation Scheme (PIC/S): The Pharmaceutical Inspection Convention and Pharmaceutical Inspection Co-Operation Scheme (jointly known as PIC/S) develop international standards between countries and pharmaceutical inspection authorities, to provide a harmonised and constructive co-operation in the field of Good Manufacturing Practices. PIC/S provides an active and constructive cooperation in the field of GMP and related areas. The purpose of PIC/S is to facilitate:

> ➤ networking between participating authorities
> ➤ maintenance of mutual confidence
> ➤ exchange of information and experience
> ➤ mutual training of GMP inspectors.

The guide consists of an introduction section along with two parts and a number of annexes.

- **Guide to Good Manufacturing Practice for Medicinal Products –** Introduction

  - o Introduction
  - o Adoption and entry into force
  - o Revision history

- **Guide to Good Manufacturing Practice for Medicinal Products -** Part I

Part I covers GMP principles for the manufacture of medicinal products

1. Quality management
2. Personnel
3. Premises and equipment
4. Documentation
5. Production
6. Quality control
7. Contract manufacture and analysis
8. Complaints and product recall
9. Self-inspection

- **Guide to Good Manufacturing Practice for Medicinal Products** - Part II
  Part II covers GMP for active substances used as starting materials

1. Introduction
2. Quality management
3. Personnel
4. Buildings and facilities
5. Process equipment
6. Documentation and records
7. Materials management
8. Production and in-process controls
9. Packaging and identification labelling of APIs and intermediates
10. Storage and distribution
11. Laboratory controls
12. Validation
13. Change control

14. Rejection and re-use of materials
15. Complaints and recalls
16. Contract manufacturers (including laboratories)
17. Agents, brokers, traders, distributors, repackers and relabellers
18. Specific guidance for APIs manufactured by cell culture / fermentation
19. APIs for use in clinical trials
20. Glossary

The annexes provide detail on specific areas of activity and are listed below:

- **Technical interpretation of PIC/S GMP guide Annex 1 -** Manufacture of sterile medicinal products

  PIC/S has published a recommendation for the technical interpretation of Annex 1 on the manufacture of sterile medicinal products.

  This recommendation summarises the interpretations an inspector adopts during an inspection of the manufacture of sterile medicinal products. It reflects the most important changes introduced in the revised Annex 1, but is not intended to address all changes in the revision.

  o Document history
  o Purpose and scope
  o Basics
  o Definitions and abbreviations
  o New texts and their interpretation

o   Revision history

- **Guide to Good Manufacturing Practice for Medicinal Products – Annexes**

  o   Annex 1 - Manufacture of sterile medicinal products
  o   Annex 2 - Manufacture of biological medicinal products for human use
  o   Annex 3 - Manufacture of radiopharmaceuticals
  o   Annex 4 - Manufacture of veterinary medicinal products other than immunologicals
  o   Annex 5 - Manufacture of immunological veterinary medical products
  o   Annex 6 - Manufacture of medicinal gases
  o   Annex 7 - Manufacture of herbal medicinal products
  o   Annex 8 - Sampling of starting and packaging materials
  o   Annex 9 - Manufacture of liquids, creams and ointments
  o   Annex 10 - Manufacture of pressurised metered dose aerosol preparations for inhalation
  o   Annex 11 - Computerised systems
  o   Annex 12 - Use of ionising radiation in the manufacture of medicinal products
  o   Annex 13 - Manufacture of investigational medicinal products
  o   Annex 14 - Manufacture of products derived from human blood or human plasma
  o   Annex 15 - Qualification and validation
  o   Annex 16 - Qualified person and batch release
  o   Annex 17 - Parametric release

- o Annex 18 - GMP guide for active pharmaceutical ingredients (This annex no longer exists)
- o Annex 19 - Reference and retention samples
- o Annex 20 - Quality risk management
- o Glossary

EudraLex - Volume 4 - Good Manufacturing Practice (GMP) Guidelines

Volume 4 of the rules governing medicinal products in the European Union contains guidance for the interpretation of the principles and guidelines of good manufacturing practices for medicinal products for human and veterinary use laid down in Commission Directives 91/356/EEC, as amended by Directive 2003/94/EC, and 91/412/EEC respectively.

EudraLex V4 is made up of the following parts:

- ➢ Introduction
- ➢ Part I - Basic requirements for medicinal products
- ➢ Part II - Basic requirements for active substances used as starting materials
- ➢ Part III - GMP related documents

Introduction

The Commission Directive 2003/94/EC, of 8 October 2003, sets out the principles and guidelines of good manufacturing practice in respect of medicinal products for human use and investigational medicinal products for human use.

Part I - Basic Requirements for Medicinal Products

Part II - Basic Requirements for Active Substances Used as Starting Materials

Basic requirements for active substances used as starting materials.

Part III - GMP Related Documents

Site Master File
Q9 Quality Risk Management
Q10 Note for Guidance on Pharmaceutical Quality System
MRA Batch Certificate

Annexes

Annex 1- Manufacture of Sterile Medicinal Products
Annex 2- Manufacture of Biological Active Substances and
Medicinal Products for Human
Annex 3- Manufacture of Radiopharmaceuticals
Annex 4- Manufacture of Veterinary Medicinal Products
Other than Immunological Veterinary Medicinal Products
Annex 5- Manufacture of Immunological Veterinary
Medicinal Products
Anne 6- Manufacture of Medicinal Gases
Annex 7- Manufacture of Herbal Medicinal Products
Annex 8- Sampling of Starting and Packaging Materials
Annex 9- Manufacture of Liquids, Creams and Ointments
Annex 10- Manufacture of Pressurised Metered Dose
Aerosol Preparations for Inhalation
Annex 11- Computerised Systems
Annex 12- Use of Ionising Radiation in the Manufacture of
Medicinal Products
Annex 13- Manufacture of Investigational Medicinal
Products
Annex 14- Manufacture of Products Derived from Human
Blood or Human Plasma

Annex 15-Qualification and Validation (in operation since 1 October 2015)

Annex 16- Certification by a Qualified Person and Batch Release

Annex 17- Parametric Release

Annex 19- Reference and Retention Samples

FDA Guidance

The FDA publishes regulations and guidance documents for industry in the Federal Register. The FDA's website also contains links to the cGMP regulations and guidance documents as well as various resources to help drug companies comply with the law. The FDA also conducts onsite audits and public outreach through presentations at national and international meetings and conferences on the subject of cGMP requirements.

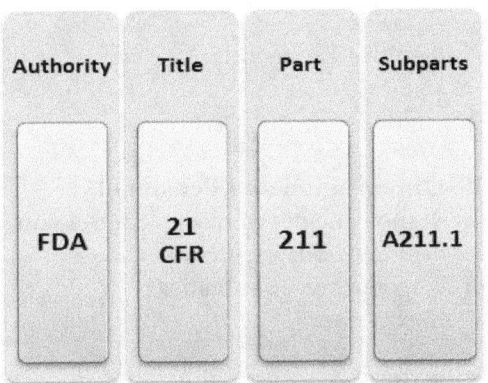

Figure 2: The FDA organises its regulations under titles. Within titles there are parts and subparts.

Pharmaceutical quality affects every American. The FDA regulates the quality of pharmaceuticals very carefully. The main regulatory standard for ensuring pharmaceutical quality is the Current Good Manufacturing Practice (CGMPs) regulation for human pharmaceuticals. Consumers expect that each batch of medicines they take will meet quality standards so that they will be safe and effective. Most people, however, are not aware of CGMPs, or how the FDA ensures that drug manufacturing processes meet these basic objectives. Recently, the FDA has announced a number of regulatory actions taken against drug manufacturers based on the lack of CGMPs. This paper discusses some facts that may be helpful in understanding how CGMPs establish the foundation for drug product quality.

PART 211   Current Good Manufacturing Practice for Finished Pharmaceuticals

Subpart A--General Provisions
§ 211.1 - Scope
§ 211.3 - Definitions

Subpart B--Organisation and Personnel
§ 211.22 - Responsibilities of quality control unit
§ 211.25 - Personnel qualifications
§ 211.28 - Personnel responsibilities
§ 211.34 - Consultants

Subpart C--Buildings and Facilities
§ 211.42 - Design and construction features
§ 211.44 - Lighting

§ 211.46 - Ventilation, air filtration, air heating and cooling
§ 211.48 - Plumbing
§ 211.50 - Sewage and refuse
§ 211.52 - Washing and toilet facilities
§ 211.56 - Sanitation
§ 211.58 - Maintenance

Subpart D--Equipment
§ 211.63 - Equipment design, size, and location
§ 211.65 - Equipment construction
§ 211.67 - Equipment cleaning and maintenance
§ 211.68 - Automatic, mechanical, and electronic equipment
§ 211.72 - Filters

Subpart E--Control of Components and Drug Product Containers and Closures
§ 211.80 - General requirements
§ 211.82 - Receipt and storage of untested components, drug product containers, and closures
§ 211.84 - Testing and approval or rejection of components, drug product containers, and closures
§ 211.86 - Use of approved components, drug product containers, and closures
§ 211.87 - Retesting of approved components, drug product containers, and closures
§ 211.89 - Rejected components, drug product containers, and closures
§ 211.94 - Drug product containers and closures

Subpart F--Production and Process Controls
§ 211.100 - Written procedures; deviations
§ 211.101 - Charge-in of components

§ 211.103 - Calculation of yield
§ 211.105 - Equipment identification
§ 211.110 - Sampling and testing of in-process materials and drug products
§ 211.111 - Time limitations on production
§ 211.113 - Control of microbiological contamination
§ 211.115 - Reprocessing

Subpart G--Packaging and Labelling Control

§ 211.122 - Materials examination and usage criteria
§ 211.125 - Labelling issuance
§ 211.130 - Packaging and labelling operations
§ 211.132 - Tamper-evident packaging requirements for over-the-counter (OTC) human drug products
§ 211.134 - Drug product inspection
§ 211.137 - Expiration dating

Subpart H--Holding and Distribution

§ 211.142 - Warehousing procedures
§ 211.150 - Distribution procedures

Subpart I--Laboratory Controls

§ 211.160 - General requirements
§ 211.165 - Testing and release for distribution
§ 211.166 - Stability testing
§ 211.167 - Special testing requirements
§ 211.170 - Reserve samples
§ 211.173 - Laboratory animals
§ 211.176 - Penicillin contamination

Subpart J--Records and Reports

§ 211.180 - General requirements
§ 211.182 - Equipment cleaning and use log

§ 211.184 - Component, drug product container, closure, and labelling records
   § 211.186 - Master production and control records
   § 211.188 - Batch production and control records
   § 211.192 - Production record review
   § 211.194 - Laboratory records
   § 211.196 - Distribution records
   § 211.198 - Complaint files

Subpart K--Returned and Salvaged Drug Products
   § 211.204 - Returned drug products
   § 211.208 - Drug product salvaging

World Health Organisation GMP Guideline Annexes

The WHO Essential Medicines and Health Products (EMP) Department works with countries to promote affordable access to quality, safe and effective medicines, vaccines, diagnostics and other medical devices. As part of this effort, the WHO publishes a number of guidance annexes that describe best practice quality requirements for specific areas within the life science industry.
List of WHO GMP Annexes:

- WHO Good Manufacturing Practices for Pharmaceutical Products: Main Principles
Annex 2, WHO Technical Report Series 986, 2014
- Active Pharmaceutical Ingredients (Bulk Drug Substances)
Annex 2, WHO Technical Report Series 957, 2010

- Active Pharmaceutical Ingredients - Bulk Drug Substances: Additional Clarifications and Explanations
- Pharmaceutical Excipients
  Annex 5, WHO Technical Report Series 885, 1999
- WHO Good Manufacturing Practices for Sterile Pharmaceutical Products
  Annex 6, WHO Technical Report Series 961, 2011
- WHO Good Manufacturing Practices for Biological Products
  Annex 3, WHO Technical Report Series 996, 2016
- WHO Good Manufacturing Practices for Blood Establishments (jointly with the Expert Committee on Biological Standardisation)
  Annex 4, WHO Technical Report Series 961, 2011
- Pharmaceutical Products Containing Hazardous Substances
  Annex 3 WHO Technical Report Series 957, 2010
- Investigational Pharmaceutical Products for Clinical Trials in Humans
  Annex 7, WHO Technical Report Series 863, 1996
- Herbal Medicinal Products
  Annex 3, WHO Technical Report Series 937, 2006
- Radiopharmaceutical Products
  Annex 3, WHO Technical Report Series 908, 2003
- Water for Pharmaceutical Use
  Annex 2, WHO Technical Report Series 970, 2012
- WHO Guidelines on Good Manufacturing Practices for Heating, Ventilation and Air-Conditioning Systems for Non-Sterile Pharmaceutical Dosage Forms
  Annex 5, WHO Technical Report Series 961, 2011

- Validation
  Annex 4, WHO Technical Report Series 937, 2006
- Guidelines on Good Manufacturing Practices:
  Validation, Appendix 7: Non-Sterile Process
  Validation
  Annex 3, WHO Technical Report Series 992, 2015

International Council for Harmonisation, ICH, GMP Guide

The International Council for Harmonisation of (Technical Requirements) for Pharmaceuticals for Human Use (ICH) brings together the regulatory authorities and pharmaceutical industry to discuss scientific and technical aspects of drug registration. Since its inception in 1990, ICH has gradually evolved, to respond to the increasingly global face of drug development.

ICH Q7 Good Manufacturing Practice Guide for Active Pharmaceutical Ingredients

Quality Management

What Is Quality?

Quality can be defined as the ability to consistently produce products meeting the same specifications time after time. Products must be safe, pure, uniform and effective. Specifications can be set down internally within a company, however, depending on the product, external specifications from regulators or standards may be required.

Patient safety is the primary focus of any pharmaceutical drug or medical device. This is the expectation of any patient or user. Secondly, the patient or user is interested in receiving an effective product. It is product specifications that ensure these criteria are accounted for.

What Is a Quality Management System?

A Quality Management System, often abbreviated to (QMS), is any system based on a collection of business processes that are primarily focused on providing safe and quality products that consistently meet customer requirements. The core themes of a QMS are outlined below:

Customer and Regulatory Focus

An understanding of the customer needs and requirements should be evident within the organisation and with the future vision of the company. The company should have an understanding of the regulatory landscape as this is subject to change over time. In turn, the company should be positioned to respond to that change.

Leadership

To truly lead, one must be accepted in the hearts and minds of those they lead. Authentic leadership pays off. A leader should foster a sense of togetherness and common vision. A leader is anyone who influences or directs people either formally or informally. We are all leaders to some extent.

Involvement

Engagement by everyone across an organisation is now recognised as being key in the successful deployment of any Quality Management System. Everyone should have a voice within the company. The saying "we are only as strong as the weakest link" is very apt here.

Systems Management

This essentially means that systems are defined and described in writing along with the appropriate responses to expected issues that arise. Effective systems management must ensure that the various systems work in support of each other and communicate effectively with one another.

Decision Making

In order to make the right decision, the person empowered to make that decision must be informed. To be correctly informed one must have the correct details and facts available. In a manufacturing environment the facts are essentially data and the analysis of data. During manufacturing or processing, data is generated as a result of monitoring and measurement of products and the related processes.

Supplier Management

Don't ruffle your suppliers' feathers. Security of supply is key in delivering products to customers or patients again and again. Raw materials or sub-components sourced from external suppliers must always be sourced at the right price and time with the emphasis on getting the best quality possible.

Continuous Improvement

For ISO 13485, continuous improvement refers to improving the effectiveness of the Quality Management System. It is harder to drive improvement of the product due to regulatory and practical requirements.

The key elements of a QMS are listed below:

Quality Policy: A company will document their commitment and approach to quality within their organisation. It usually sets out how they plan to achieve a high and consistent standard of quality. It should in some way speak to the customer or end user.

Quality Objectives: Quality objectives can be documented in a quality plan at site or organisational level. An effective way of defining quality objectives is by using SMART method. SMART stands for **S**pecific, **M**easurable, **A**chievable, **R**ealistic and **T**imely.

Quality Manual: An in-house guidance document to provide a framework for achieving the quality objectives.

Organisational Structure and Responsibilities: Organisational charts can be used to map out the company structure. Roles and responsibilities can be documented in site quality plans, job descriptions and Standard Operating Procedures.

Data Management: A coherent approach to the provision, storage and maintenance of data.

Processes: Processes are defined and documented.

Resources: Resources must be properly understood, allocated and linked across the organisation.

Product Quality and Customer Satisfaction: The proper management and investigation of complaints is important to reduce future instances from reoccurring. Continual engagement with the end user or customer is critical.

Continuous Improvement Including Corrective and Preventive Action: Where continuous improvement projects and initiatives are encouraged and supported. The application of a CAPA system to ensure quality is maintained and consistent.

Maintenance: A Preventative Maintenance schedule is in place and managed accordingly.

Sustainability: All work practices are sustainable and consistent throughout the life-cycle of processes and products.

Auditing: Systems are auditable and maintained to allow internal or external review and audit.

Engineering Change Control: Where changes are required to validated processes or equipment, changes are managed and introduced under change control.

A common acronym used to highlight the aims of Good Manufacturing Practices (GMP) is SPUE which stands for Safe-Pure-Uniform-Effective. This definition is particularly suited to pharmaceutical products as the chemicals and drugs used need to be pure and free of contaminants. Furthermore, they need to be uniform, meaning they will have the same constituents from tablet to tablet and batch to batch. A description of each word is shown below:

SAFE- the product has the right ingredients if it is a drug product. It is packaged as intended and correctly labelled in order to provide identification and safe use.

PURE- it is free of contaminants, foreign matter, chemicals and harmful microbes.

UNIFORM- The product is manufactured consistently and will have the same quality between batches manufactured on different days.

EFFECTIVE- Ultimately, the product must be effective in treating the medical condition. To be effective, it requires the correct ingredients, the correct amount of ingredients and correct packaging to maintain the product stability over time.

The basic concepts of Quality Management, Good Manufacturing Practice and Quality Risk Management are inter-related. They are described here in order to emphasise their relationships and their fundamental importance to the production and control of medicinal products.

Figure: Elements of a Quality System

A Pharmaceutical Quality System must be appropriate for the manufacture of medicinal products and incorporate management controls that ensure patient safety.

1) Product realisation is the process of identifying market opportunities and user needs and bringing them forward through a design planning stage that will result in a new product. A systematic approach to introducing new products is important to ensure consistency.

2) Product and process knowledge must be managed throughout all stages of a product from design and development to ultimate retirement. Training, formal documentation and design specifications all contribute to the knowledge pool.

3) Medicinal products are designed and developed in a way that takes account of the requirements of Good Manufacturing Practice.

4) Both production and control operations are clearly specified and Good
Manufacturing Practice is adopted.

5) Managerial responsibilities are clearly specified.

6) Arrangements are made for the manufacture, supply and use of the correct starting and packaging materials, the selection and monitoring of suppliers and for verifying that each delivery is from the approved supply chain. (e.g. supplier approval processes).

7) Systems of control are established and maintained by developing and using effective monitoring and control systems for both process performance and product quality. Metrics help ensure processes are in control and product quality is being maintained.

8) Continuous improvement is facilitated through the implementation of quality improvements appropriate to the current level of process and product knowledge.

9) Arrangements are in place for the prospective evaluation of planned changes and their approval prior to implementation taking into account regulatory notification and approval where required.

10) Formal tools such as root cause analysis are applied during the investigation of deviations, suspected product defects and other quality issues. This can be determined using Quality Risk Management principles.

Medicinal products should not be sold or supplied before a qualified person has certified that each production batch has been produced and controlled in accordance with the requirements of the competent authority and any other regulations relevant to the production, control and release of medicinal products. There should also be a process for self-inspection and/or quality audits that examines regularly the effectiveness and applicability of the Pharmaceutical Quality System.

Quality Control

Quality Control is that part of Good Manufacturing Practice which is concerned with sampling, specifications and testing, and with the organisation, documentation and release procedures which ensure that the necessary and relevant tests are actually carried out and that materials are not released for use, nor products released for sale or supply, until their quality has been judged to be satisfactory. The basic requirements of Quality Control are that:

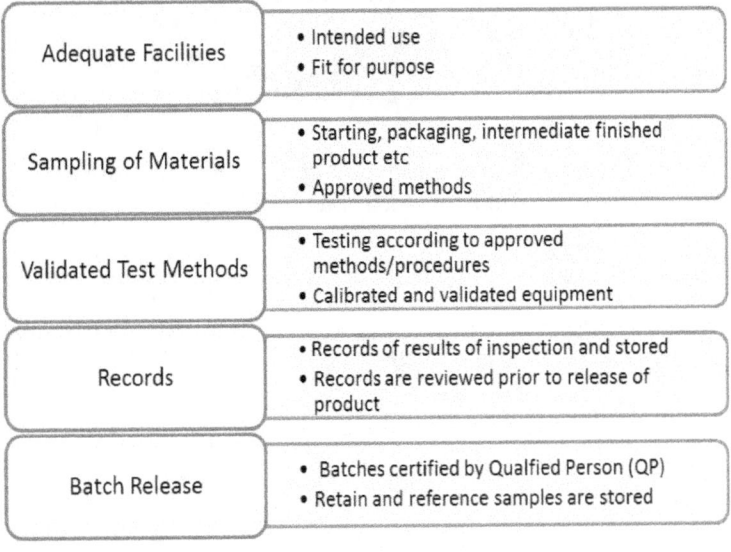

Figure: Control elements within a Quality Management System

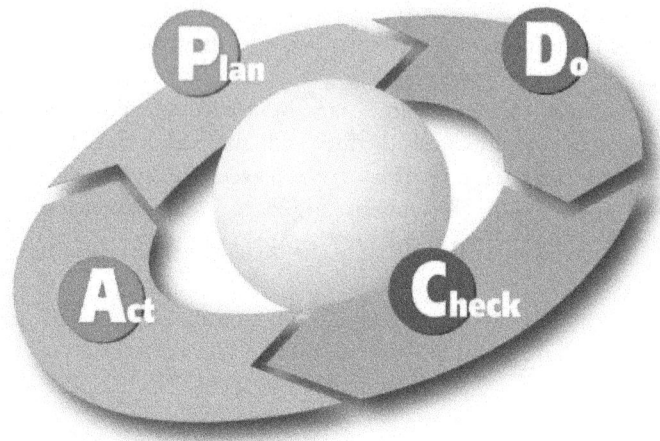

Figure: Plan, Do, Check Act, PDCA- continuous improvement methodology.

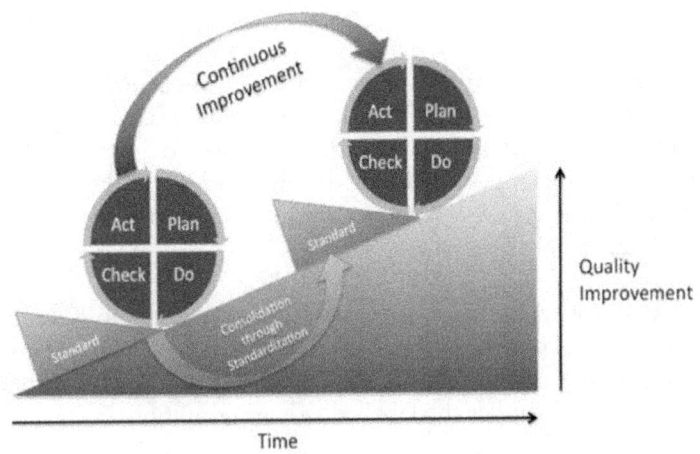

Figure: The PDCA implemented continuously over time

Ongoing Stability

After marketing, the stability of the medicinal product should be monitored according to a continuous appropriate programme that will permit the detection of any stability issue (e.g. changes in levels of impurities or dissolution profile) associated with the formulation in the marketed package.

The purpose of the ongoing stability programme is to monitor the product over its shelf life and to determine that the product remains, and can be expected to remain, within specifications under the labelled storage conditions.

The protocol for an ongoing stability programme should extend to the end of the shelf life period and should include at a minimum, the following:

(i) Number of batch(es) per strength and different batch sizes, if applicable
(ii) Physical, chemical, microbiological and biological test methods
(iii) Acceptance criteria

(v) Description of the container closure system(s) or packaging
(vi) Testing intervals (time points)
(vii) Description of the conditions of storage

A summary of all the data generated, including any interim conclusions on the programme, should be written and maintained. This summary should be subjected to periodic review.

Personnel

Personnel are central to the application of CGMP and compliance to regulations. At every level throughout an organisation, people interact with materials, equipment and processes in order to deliver products to the market and patient. Personnel must therefore be suitably qualified and equipped to carry out their responsibilities effectively.

Provisions in guidance and regulations are therefore made for personnel in a Quality Management System. Despite advances in automation and computerised systems, people are centrally involved in day-to-day decisions. For this reason there must be sufficient and suitably qualified personnel to carry out all the tasks. Individual responsibilities should be clearly defined and understood by the persons concerned and recorded formally in procedures and job descriptions.

General

It may be an obvious point; however, manufacturers must ensure an adequate number of personnel with the necessary qualifications and practical experience are resourced to manufacturing. Having a broad base of people with the experience, knowledge and skills reduces the risk of quality issues. Responsibilities placed on any one individual should not be so extensive as to present any risk to quality.

Personnel should have specific duties recorded in written descriptions and adequate authority to carry out their responsibilities. Their duties may be delegated to designated deputies with a satisfactory level of qualifications.

Personal Hygiene

All personnel should be trained in the practices of personal hygiene. A high level of personal hygiene should be observed by all those concerned with manufacturing processes. Personnel should be instructed to wash their hands before entering production areas. Signage should be in place along with hand washing facilities. Hand washing demonstrations and training should be provided by a suitably qualified QC analyst or microbiologist. Any person experiencing an illness or exhibiting open lesions or wounds that may adversely affect the quality of products should not be allowed to handle starting materials, packaging materials, in-process materials or medicines until the condition is no longer a risk to quality or patient safety.

Direct contact should be avoided between the operator's hands and starting materials, primary packaging materials and intermediate or bulk product.

Buildings and Facilities

Facilities and utilities qualifications are typically prerequisites to the validation of manufacturing equipment and systems. Much of the activity that deals with establishing a facility or building that is *fit for purpose* and managed under the broad heading of commissioning and qualification (C&Q). The terms C&Q are often used interchangeably and in practice some overlap in activity is expected. Commissioning can be defined as the planned, documented, and managed engineering approach to the start-up and handover of facilities, systems and equipment to the end-user. It must deliver a safe and functional environment that meets the predefined design and user requirements.

In strict terms, qualification is more concerned with the confirmation and documentation showing that equipment and systems are properly installed and functional. Qualification forms part of validation but the individual qualification steps do not equal a validated process. The establishment of a User Requirements Specification (URS) and detailed design specifications ensure that the building or facility will meet end users' needs and that it is fit for the intended purpose.

It also provides a level of protection to the contracting company responsible for the project or facility construction. Post-URS approval requires an approved Design Qualification (DQ). This provides verification and a documented record that the proposed design is suitable for the intended purpose. Further verification including IQ/OP/PQ should be applied as required based on the system impact and criticality of facilities/utilities.

The design and layout of any manufacturing area should facilitate the effective flow of materials. This is a fundamental requirement no matter what the industry, e.g. medical devices, pharmaceuticals, bio pharmaceuticals and even non-regulated engineering companies that assemble, machine or fabricate products. However, manufacturing medicinal products that are required to be sterile imposes a greater level of control and thought. With regard to aseptic processing facilities, material flows do not only require efficient and effective flow of materials; the activity should support the requirements of aseptic processing while minimising any risk of contamination. Identifying critical processing zones is an important step in ensuring the right building design and controls are implemented. Isolators and aseptic filling require the highest classification with strict environmental controls. Secondary packaging operations such as cartonning are often completed in areas controlled and operated to a lower classification.

Design and layout of facilities should:

> Maintain microbiological integrity of the identified critical processing zones
> Prevent or minimise contamination from outside critical processing zones
> Control the flow of materials by restricting access to trained and authorised personnel

Material Transfer

Material transfer from the outside of cleanrooms to the inside is completed via material air locks or hatches. Material air locks and hatches ensure that there is clear separation between controlled clean areas and less clean areas. Many suppliers provide products that are double bagged. This provides an added level of control when transferring materials. The outer bag can be removed within the air lock thus providing a clean inner product. Material air locks also allow the sanitisation of products. Tools and other items must be clean and dirt free.

Material Air lock Considerations:

- ➢ Interlocked doors
- ➢ Access control
- ➢ Sanitation/ Cleaning procedure
- ➢ Double or triple-bagged products
- ➢ Dedicated trolley for air locks

Disinfection and Cleaning Agents

When materials are being transferred via an air lock, consideration must be given to the status of materials and products. As a rule, no cardboard or unnecessary paper should enter a cleanroom. Wooden pallets are not acceptable as they can carry dirt and microorganisms, and wood cannot be sanitised due to its porous nature. Soft fabric cases often used to carry tools should also be avoided as the material can carry dirt and grease. Cleaning and disinfecting agents should be tested and approved prior to their use onsite. The choice of agents should be backed up with studies that demonstrate the effectiveness of disinfectants and cleaning agents.

Gown-Up Areas

Gowning rooms are designed in order to minimise contamination and facilitate the orderly change over from street clothes to scrubs and/or gowns. Hand washing facilities help reduce the risk of humans carrying unwanted microorganisms into the aseptic processing area. The design of the room should result in clear separation between the less clean side and the clean side. This can be achieved with a step-over segregating the two areas.

Other features of gowning rooms should include:

- ➢ Storage lockers for street clothes
- ➢ Gown and garment storage
- ➢ Body length mirrors
- ➢ Hand washing /drying and disinfection facilities

GMP Zoning

Selecting a suitable classification for a room or manufacturing facility depends on several factors. Firstly, it can be said that sterile products require a more stringent set of criteria than non-sterile products. However, there is an extensive range of products and medical devices that are sterile but are used in different ways and consist of different materials and technology. Some sterile products are single use only and used for short-term purposes and then disposed of. Other sterile products are used subcutaneously for longer periods or even require implantation. Therefore, the design of a facility along with its HVAC specification must be appropriate to the product being manufactured. High-risk products require greater control. The goal of facilities and HVAC systems is to minimise contamination and the associated risks. Using a sterile versus non-sterile rule of thumb is not adequate when classifying a room or facility. Standards including EN ISO 14644-1 and guidelines such as EU cGMP Guidelines EudraLex volume 4 Annex 1 (2008) should be consulted in order to fully understand the requirements of each ISO classification and grade of room.

ISO classifications do not specify room occupancy states but when a designation is applied, the occupancy state must be stated in the relevant documentation or procedure. The most relevant European Guideline (Annex 1 of the EU cGMP Guideline) lists four classification grades and their associated particulate limits in the 'at rest' and 'in operation' conditions. In general, for the sterile and non-sterile products, similar classes are applied, but in non-sterile production the producer could assign their classes, having similar particulate concentration, temperature, pressure etc. but lower air-change rate could be used.

Types of Contamination:

- cross contamination (of a product/material with another product/material)
- non-microbial particulate contamination (non-viable particles)
- biological/microbiological contamination (viable particles/micro-organisms)

Factors Influencing Contamination Cleanliness Levels in the Manufacturing Processes:

- process
- air cleanliness
- personnel hygiene and clothing
- work practices
- material design (material of construction, surface finishes, room finishes, equipment, open system/enclosed system, utensils etc.)
- material cleanliness

Room Air Classification

Environmental Grade A (Aseptic)

Grade A is reserved for critical processes in manufacturing sterile products, product components or product contact. This is generally achieved using isolator technology which maintains a barrier to the background environment or surrounding room.

Grade A Operations include:

- Aseptic processing of sterile ingredients
- Filling of sterile products not for terminal sterilisation
- Stopper insertion
- Crimp capping

Environmental Grade B

Grade B is used for supportive work for aseptic processing corresponding to ISO 14644 (Part 1) Class 5 ("at rest") and Class 7 (when "in operation"). Grade B areas typically serve as the background environment of Grade A areas for aseptic processing.

Environmental Grade C

Suitable for non-critical processing steps, Grade C corresponds to ISO 14644 Part 1 Class 7 ("at rest") and Class 8 ("in operation"). Grade C operations include:

- Clean side of material air locks and gowning rooms
- Filling of products that are to be terminally sterilised

Environmental Grade D

Grade D at least corresponds to ISO 14644 Part 1 Class 8 ("at rest" / no definition for "in operation").

- Clean section of material air locks and final compartments of gowning rooms

➢ Dispensing of raw materials and excipients and preparation of solutions for sterile products to be sterile filtered and terminally sterilised

➢ Background environment for transfer and crimp capping of stoppered containers with sterile products

Compliance Tests for GMP Zones

Test	Requirements
Particle count test	Test covers verification of cleanliness. Dust particle counts to be carried out and result printed. The number of readings and positions of tests should be defined in accordance with ISO 14644-1 Annex B5
Air pressure difference	This test is used to verify non cross-contamination. Log of pressure differential readings to be produced or critical plants should be logged daily, preferably continuously. A 15 Pa pressure differential between different zones is recommended. Refer to ISO 14644-3 Annex B5
Airflow volume	To verify air change rates. Airflow readings for supply air and return air grilles to be measured and air change rates to be calculated. Refer to ISO 14644-3 Annex B13

Airflow velocity	To verify unidirectional flow or containment conditions. Air velocities for containment systems and unidirectional flow protection systems to be measured. Refer to ISO 14644-3 Annex B4
Filter leakage tests	To verify filter integrity. Filter penetration tests to be carried out by a competent person to demonstrate filter media, filter seal and filter frame integrity. Only required on HEPA filters. Refer to ISO 14644-3 Annex B6
Containment leakage	To verify absence of cross-contamination. Demonstrate that contaminant is maintained within a room by means of: • airflow direction smoke tests • room air pressures. Refer to ISO 14644-3 Annex B4
Recovery	To verify clean-up time. Test to establish time that a cleanroom takes to recover from a contaminated condition to the specified cleanroom condition. Should not take more than 15 minutes. Refer to ISO 14644-3 Annex B13
Airflow visualisation	To verify required airflow patterns. Tests to demonstrate air flows: • from clean to dirty areas

	• do not cause cross-contamination • uniformly from unidirectional airflow units Demonstrated by actual or video-taped smoke tests. Refer to ISO 14644-3 Annex B7

Clean Room Design Considerations

Air Handling Unit (AHU) -Air Intake Quality

All locations on earth except latitudes near the equator experience seasonal temperature changes. The changes are a consequence of Earth's orbital motion about the sun, coupled with the tilt of its axis of rotation with respect to its orbital plane. Design criteria should be based on published temperature data. The HVAC system design should consider the following:

Standard Operating Conditions: These are climatic conditions against which the systems must be designed to operate, control, and maintain required conditions. (These may be based on published data, which are only exceeded 2.5% or 1% of the time).

Extreme Operating Conditions: These are climatic conditions against which the systems must be designed to operate, without manual intervention, and without damage to the systems or the facility. Based on product / process risk assessments, extreme or standard conditions shall be used for HVAC design for dedicated areas.

Location

Based on the building layout, footprint and design intent, a suitable and adequate space must be identified for HVAC location. This must include provision of chilled water, heating systems, ducts and drainage. HVAC plants must be accommodated in designated HVAC plant rooms or interstitial areas.

Air Intake

During the design phase, the air intake locations should be selected to ensure air is in the best environmental condition. The below considerations help to achieve a strong starting point:

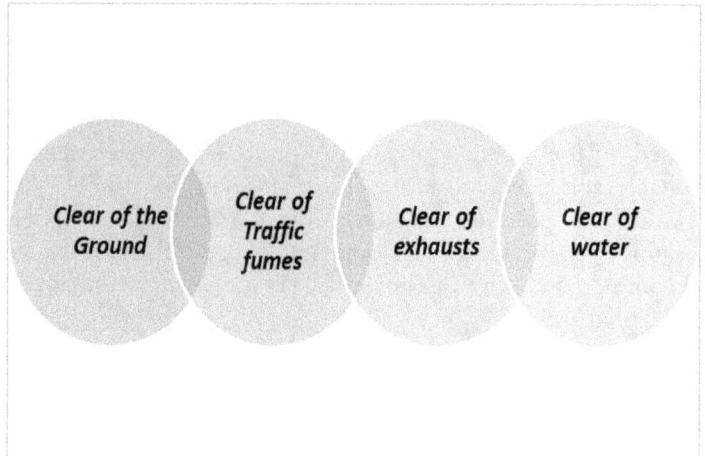

Figure : Air intake considerations

Thermal Load

Thermal load can be defined as the amount of heat energy to be removed from an inner environment by equipment (HVAC) used to maintain that environment at the design temperature when worst case external temperature(s) are being experienced. The thermal load requirement should be calculated for the following:

➢ Max summer conditions
➢ Minimum winter conditions
➢ High rainfall
➢ Standard operation
➢ Extreme operating conditions

Room Recovery Time

Room recovery time to return to the required pressure
differential and cleanliness
Specification should be minimised.

Dust, Vapour, or Fume Control

Highlight areas requiring dust, vapour, gas and/or fume
control on the room data sheet. These areas must be
controlled to remove the possibility of product
contamination and to ensure the safety of the operator and
environment. Areas requiring 100% fresh air or extraction to
atmosphere may require greater airflow or other measures
within the room to maintain environmental conditions.

In order to meet the appropriate level of cleanliness, HVAC
systems require sufficient filtration to provide "clean" air to
prevent contamination of the product. Pre-filters and main
filters are normally suitable for most operations; however,
HEPA filters are required to prevent particulate or microbial
contamination for higher-classification areas

Air Change Rates

The air change rates for each room must be calculated to be sufficient for clean-up to achieve specified particulate conditions "at rest" in static conditions after a maximum of 20 minutes from completion of operations. The actual air change rate must be chosen to satisfy the most stringent requirements including GMP, GLP, heat gain, ventilation requirements and/or occupancy, including an appropriate safety factor.

The air change rate must be optimised for energy savings; however, specific attention must be paid to air locks where a greater air change rate must be applied. Air changes can be reduced (e.g. setback modes) in some circumstances ("at rest" mode, with no production activity and no personnel interventions).

Room Environmental Conditions

Other environmental conditions to be controlled, such as temperature and relative humidity, depend on the product and nature of the operations carried out in those areas. These parameters should not interfere with the defined cleanliness standard.

Temperature Requirement

The normal operating temperature requirement for each classification .Temperature and humidity must be appropriate to the product and process. Consideration should be made for specific product and process requirements.

Humidity Requirement

The normal operating humidity requirement.

Particulate Levels

Particulate levels are specifically defined for each room classification "at rest" and "in operation". The levels are controlled though air filtration, facility design, gowning requirements, and decontamination

Room Exhaust

Where there is a risk of active compounds being present in extracted air, filters should be fitted, preferably in the room, to prevent contamination of ductwork and the environment. The filters must be selected based on the particle size distribution of the products to be handled.

HVAC System Design

The HVAC system must be appropriately selected using the specific design requirements as outlined above. The system must be able to provide clean, conditioned air to the specified areas to meet all of the quality requirements. The most important precursor to HVAC design is the comprehensive definition of the function and performance required followed by the selection of an appropriate system. A poor selection can lead to unnecessarily high-energy consumption, and operational deficiencies.

All-air systems rely on the movement of large quantities of air through a central air handling unit to control room conditions, as well as provide for ventilation requirements.

They have the advantage of being relatively simple with most of the unit situated in one location; however, they are very space consuming. All-air systems tend to be relatively inflexible and not ideal for areas that are likely to need environmental alteration on a regular basis.

These HVAC systems are used for areas that have a lot of small zones, each with slightly different thermal loads but which requires constant ventilation. These systems can have poor energy efficiency if a lot of reheat is required. These are typically used in large manufacturing areas, and laboratories with many small rooms.

Dust Extraction and Collection

It is essential to capture dust as close as possible to the point of generation without affecting the process. In most cases dust capture should be within 100mm from the point of release. Air velocity is the key parameter in dust capture.
Pharmaceutical and chemical applications have specific collection requirements as any dust build-up in the system is likely to be of a pharmacologically active nature, sensitising, toxic and/or corrosive. It is vital to maintain transport velocities and minimise any potential for cross contamination.

A typical system should have a minimum transport velocity of 18 m/s, but this may need to be higher if heavy particles are to be collected. This velocity must be maintained throughout the system to prevent dust from dropping out in the ducts.

The dust collection must be configured with the hazardous nature of the dust in mind. A clearly defined disposal procedure for the collected dust (e.g. bag-in / bag-out system for filter and dust bin) needs to be understood at the design stage. HVAC unit shall meet EN 1886 and EN 13053 requirements.

Fans

Certified performance curves are required to verify correct fan operation. Fans that may be subjected to high temperatures, humidity, corrosive fumes or other hazardous atmospheres should be constructed using non-reactive, non-corrosive, suitable and approved materials (such as epoxy painting). Whenever H2O2 or other disinfection application is planned, material compatibility certificates shall be supplied by the vendor.

Fans must be selected to supply the design volume, taking into account the assumption that filters are half clogged, except for the terminal filter which shall be considered to be fully clogged according to EN 13053. If the terminal filter is HEPA, clogging shall be considered according to EN 1822 and the target volume is 80% of the given maximum clogged specified value.

Filtration

Face-fitting filters shall be used in all cases, as slide-in filter elements never give a good seal. The installation must be such that the airflow pushes the filter against the seal. The face velocity across the filter section shall not exceed 2 m/s. For ventilation and air conditioning applications, two minimum filtration stages are required. For certain applications, return air filtration will be required to contain highly active materials (e.g. viruses or potent compounds). Normally, these filters should be changed from the room side. However, since those filters must be integrity tested, it is recommended to place one filter in the main return duct before the exhaust fan and design return duct network, in order to ensure tightness of the duct between the room and the filter (bag-in / bag-out filter change systems should be provided for BSL-3 areas). In case of live biological agent biocontainment, decontamination up to the filter must be proven. The grade of filter and technical solution must be selected based on the product particle size distribution and occupational exposure band (OEB) level.

HEPA filters and Dehumidification

For most HVAC applications, dehumidification is best achieved by the use of cooling coils. It should be noted that dehumidification is a very high consumer of energy and should only be used if there is a real process need. When areas are not in use, the dehumidifier should be turned off, if possible.

When room humidity must be maintained below 50% during warm weather, an absorption dryer may be necessary unless the room temperature can be increased within specification to compensate.

Normal practice is to optimise size and efficiency of the absorption dryer by first removing as much moisture from the air as possible by cooling. The design of absorption dryers is normally based on a slowly rotating desiccant wheel.

Air is passed through the wheel and dried by the desiccant coating (guidance: lithium chloride especially if the wheel is not used frequently and silica gel if used permanently and with low humidity target). It is not normally necessary to size a dryer to handle the entire air volume. Drying a proportion of air and re-mixing to achieve the desired moisture content is usually sufficient.

Air humidification may be necessary during cold weather when introducing fresh air to spaces that require humidity control. When air humidification is necessary, humidifiers should be selected on the following basis:

> ➤ direct steam injection using steam
> ➤ direct steam injection using self-generative electric or gas steam humidifier.

Humidifiers should be located before the fan and the final filter which will remove any particulate generated. At least 300 mm clearance should be allowed upstream and 1 m downstream between humidifier manifolds and coils, attenuators etc. (general recommendation to be confirmed through calculation note provided by the vendor). A single manifold or multiple manifolds in parallel may be used to meet the humidification requirements as per manufacturer's recommendations.

Sound Attenuators

Sound attenuators should be provided as necessary, to achieve the specified noise levels within occupied spaces. To minimise external noise nuisance, assessment can confirm the necessity to use acoustic media (enveloped in polyester film), that is inert and corrosion-resistant at normal operating conditions. Material quality shall be equivalent to that specified for HVAC unit or ducts.

Sound attenuators should be installed in the air handling unit or ductwork. The use of sound attenuators in the air supply and air return should be based on requirements for fresh air inlet and air exhaust, and according to external noise levels that might need to be maintained at or below the ambient site noise levels.

Dampers

The provision of sufficient dampers is essential for proper control. To minimise noise transmission into the room, these should be mounted as far as possible from the diffuser.

Carefully evaluate the space-by-space pressure control that will be used in the design. Static pressure control via hard balance or dynamic control via air terminal control units are both appropriate. Consideration should be given to the overall project size, the complexity of the facility and the project budget.

Automatic volume controllers are recommended for regulating air volume independently of supply pressure. They can be selected for constant volume, variable volume or dual duct mixing applications. Automatic low-leakage fresh air and exhaust air shutoff dampers are strongly recommended to isolate the HVAC network. Fresh air dampers shall be Class 3 minimum (maximum leakage preventing coil freezing). Whenever fumigation is performed shutoff damper shall ensure Class 4 leakage rate. Where dampers are required to provide modulating control of airflow, they must be selected to provide an appropriate level of control authority. This will normally mean a damper smaller than the duct size.

Heating and Cooling

Heating mode: Low pressure hot water (LPHW) is the preferred heating medium for HVAC applications and should be used whenever practicable. Electrical heating should be avoided due to fire risk and should be limited to low power coil and in locations where no other energies are available. Hazard operability analysis (HAZOP) must be conducted if electrical heating is being considered. Cooling mode: Chilled water is the preferred cooling medium for HVAC applications and should be used whenever practicable.

The direct expansion of refrigerant in coils is an acceptable method of cooling, particularly on small isolated plants, or where lower temperatures are needed for dehumidification or for cold room. This system, however, does not normally give close control. Direct expansion coils should only be used with extreme care on variable air volume systems (if speed driver available on compressors).

Heating Coils

The face velocity of air across heating coils should not exceed 2 m/s. Coils should be made of material suitable for applicable constraints. Drains shall be located outside the casing of the HVAC unit. Coils shall be removable.

Cooling Coils

Cooling coils have been identified as potential sources of microbial contamination; therefore, careful design is required to prevent water carryover and to ensure that drain pans do not retain water. Double tube, non-welded units are recommended. The face velocity of air across cooling coils should not exceed 2 m/s. Where necessary, stainless steel or plastic eliminator blades should be provided to prevent any moisture carryover. Where provided, these must be removable for cleaning.

Ductwork

For most applications, galvanised steel ductwork will be the most appropriate form of construction; however, stainless steel or plastic construction may be necessary where there is a higher risk of corrosion due to moisture or fumes (exhaust ducts usually). Where operating pressures above 2,000 Pa are necessary, fully welded construction is recommended. For contained ducts (e.g., exhaust duct before bag-in / bag-out filter), air tightness Class C shall be followed (EN 12237). For BSL-3, fully welded construction should be considered.

Generally ductwork should be constructed to an appropriate local standard, suitable for the maximum design pressure (positive or negative), such as those published by Sheet Metal and Air Conditioning Contractors' National Association (SMACNA) in the USA, Building and Engineering Services Association (B&ES) in the UK

Where flexible connections are proposed these must be designed for the same pressure as the ductwork. Solid ducted connections are preferred for final connections to terminal HEPA filter housings. For applications where flexible connections to diffusers are used, these should be no longer than 500 mm and nominally straight.

Special consideration must be given to fume extract ducts where these pass through fire barriers. Using fire dampers should be avoided where the loss of extraction could make a fire situation worse. An alternative design, such as the use of fire-rated ductwork, may be necessary in these cases. A thorough risk assessment must be conducted.

Parameter	Description
Temperature	The HVAC must be capable of

Simple Representation of HVAC system

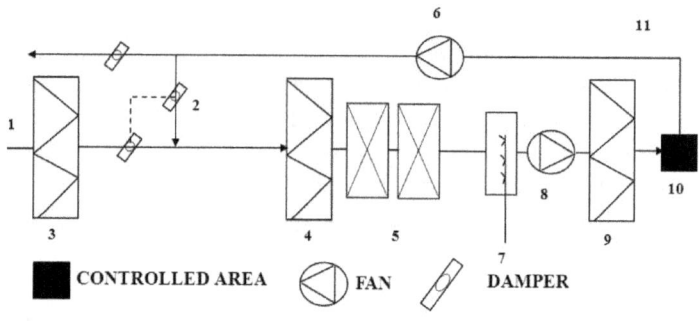

Figure: Simple HVAC diagram

Position	Description
1	Fresh air intake (°C, %RH, flow rate)
2	Dampers
3	Filter creating a differential pressure
4	Filter creating a differential pressure
5	Control valves for cooling fluid
6	Exhaust fan
7	Steam flow rate
8	Supply fan
9	Filter creating a differential pressure
10	Controlled room/ area
11	Extraction

	operating over a range of temperatures and accurate to a tolerance. Temperature probes/detectors must be placed at various points to provide feedback and control.
Relative Humidity	Relative humidity must be monitored continuously. Typically, humidity sensors should be effective over an operational range (e.g. 5-95% R.H.) Accuracy should also be no less than ±3%
Air flow/ Air Pressure	Air flow is proportional to the square root of the differential pressure.
Dampers	Dampers are used to control inlet and outlet airflow.
Valves	Ball valves or globe valves control the flow of air. Valves are designed with specific safety features to meet the intended use (e.g. CLOSED without energy supply-cooling valve, OPEN without energy supply-heating valve.

An environmental monitoring programme is required for GMP controlled areas. The purpose of such programmes is to document, define and describe parameters to be monitored, monitoring frequency and methods. Environmental monitoring is a regulatory requirement. It also demonstrates that the GMP areas are been controlled and are fit for purpose.

Key Requirements of Environmental Monitoring

Identification and classification of environmental areas that require monitoring

Test methods and sampling procedures

Defined testing frequencies

Sample locations based on Risk

Microbial monitoring of personnel

Monitoring of non viable particles

Monitoring of temperature, relative humidity and differential pressures

Defined alert and action levels for each environmental area

Trending of Enviromental data

Change Control

Figure: Key requirements of environmental monitoring

Other parameters such as those controlled by the HVAC system (air changes/hour etc.) should also be verified according to a defined schedule.

Grade A, B and C

- ➢ Viable and non-viable particles monitored under operational conditions
- ➢ Risk-based approach to sampling points that represent high risk/critical positions

Grade D

- ➢ Non-viable particles must be measured at-rest conditions
- ➢ Viable particles measured under operational conditions

Documentation and Records

Good documentation is an essential part of the quality assurance system and is relevant across all departments and functions within a manufacturing company. Controlled documents define the specifications and procedures for all materials and methods of manufacture and control strategies.

General Requirements

- ➢ Documents should be designed, prepared, reviewed, approved and distributed in accordance with approved processes.
- ➢ Documents should comply with relevant parts of the manufacturing and marketing authorisations.

> ➤ Documents should be approved, signed and dated by the appropriate responsible persons. No document should be changed without authorisation and approval.
> ➤ Documents should have unambiguous contents: the title, nature and purpose should be clearly stated. They should be laid out in an orderly fashion and be easy to check.
> ➤ Documents must not allow any error to be introduced through the reproduction process.

Good Documentation Practices

This section provides an easy-to-understand guide to the subject of Good Documentation Practices. Good Documentation Practices (commonly abbreviated to GDP or GDocP) is a term used to describe standards by which documents are created, modified and maintained. The need for GDP is driven by the general requirement of GMP (Good Manufacturing Practices).

GDP is a practical skill that is required within the life science sector (medical device, pharmaceutical and so on). It can be broadly divided into two streams; GDocP practices and how they apply to electronic document and secondly, GDocP for handwritten entries including initial and dating and recording of data and test results by hand. GDocP is fundamental in achieving compliance to Good Manufacturing Practices (GMP). It is required in the U.S. by the FDA's Code of Federal Regulations and in Europe by the governing body EudraLex. If GDocP is not practiced it jeopardises the integrity of data and written records. This can lead to the falsification of data which is a serious regulatory offence. Admittedly, implementing and maintaining GDP takes time, effort and resources, however, there are some benefits that come with it. Most importantly, Good Documentation Practice is an expected practice in regulated industry as trust and ethics are fundamental to business.

It is important to maintain accurate, complete, up-to-date and consistent information to ensure patient safety and reduce any potential risk to patients. Practicing GDP equally helps to reduce observations raised on inadequate documentation practices at times of audit by regulated bodies such as the FDA. It helps to improve communication and efficiency within companies. If GDP is not followed it can call into question other processes and procedures within a company.

Documentation Creation

The principles of GDP should be applied at the document creation stage. As most people are familiar with softcopy or electronic documents, some of these points are obvious but nonetheless need to be made. All documents should be electronically written and not handwritten except for execution of protocols, test results and adding entries. Documents that are approved controlled should be:

- Accurate and free from errors
- Have revision or be version controlled
- Should have an effective date or date of release

Approval of Documents

Document approval must be completed by trained and appropriately experienced personnel. Often companies will use an approval matrix which explains which people are required to approve each document. For example, an EHS (Environment Health and Safety) officer would be required to approve a risk assessment.

Signatures

A signature on any document is legally binding so remember to read and understand what is being signed for. Every signature should also include the date in the correct format. If a signature appears within the same document alongside initials, substituting a full signature with initials and date is generally acceptable. This practice is common when large documents are being completed.

Date and Time Format

A standardised approach to dates and times is important especially within large global organisations. For instance, in the USA, the norm is to place the month before the date, whereas in Ireland and Great Britain it is common to write the day of the month followed by the month. Most companies would define their date and time format in an SOP or procedure.

The date and time format can also be configured in Word documents and Excel worksheets to align with a companies preferred date and time format.

Handwritten Entries

When a handwritten entry is required such as a signature or a test result, indelible ink must be used. Many companies will have an SOP or procedure that states the specific ink colour required. If an entry of a test result or test data isn't completed at the time of execution, this constitutes a late entry. Backdating an entry or signature is forbidden. Always use the correct and current date.

How Are Mistakes Corrected?

This is a critical area of GDP. Failure to follow the requirements of GDP when correcting mistakes is the most common failure when it comes to documentation in industry. The method of correcting mistakes using GDP allows for a person looking at the document for the first time to clearly see the original entry and the corrected entry. This maintains the integrity of the document. In order to identify the changes and corrections, certain rules must be followed. No overwriting is allowed and white-out or Tipp-Ex is unacceptable.

Accuracy

Accuracy of information provided in documents is critical in the life science industry. As the end user is a patient, inaccurate records or documents could cause serious injury or death. Controlled documents are also legal documents and could be called upon if recalls, litigation or investigations arise.

Many documents used in the manufacture of medical devices are designed to record information or test results. These test results are then used to disposition (pass or fail) batches of product. Inaccurate information could risk the release and distribution of defective product. This has a potential impact on both the business and the patient or user.

Blank Spaces or Blank Fields

On completion of a document such as a logbook or record, no blanks spaces should be left unfilled. This is to avoid late entries and also to prevent confusion. Blank spaces or blank fields should have a diagonal line drawn neatly across the space, the letters "N/A" written and the entry signed and dated. If the reason for "N/A" is not evident then it is wise to include an explanatory note or sentence.

Data Transcription

Transcribing is the process of transferring data from one source to another. This is often required when raw data is involved. When data is in raw format it may need to be entered into a Microsoft Excel sheet. When transcribing data, remember that all original raw data must be stored in case it is needed at a future date. After the data is transcribed it must be verified by a second person to check for any errors or omissions.

Revision Control

Controlled documents should always have a version number or revision number electronically on each page of the document. This is similar to books which always list what edition they are e.g. first edition or second edition. Revision control is a key element of Quality Management Systems in place in regulated industries. As the need for changes in the document arises, the controlled document can be amended/updated. With each update the version number revises also. Some companies will use alphabetic revision control and to a lesser extent numeric revision control (Version A, Version B or Version 01, Version 02).

Management of Attachments

Attachments to controlled documents can include training records, data sheets, lab results and so on. It is important that attachments are identified for traceability purposes. If the attached becomes detached from the main document, the attachment should be identifiable. It is best practice to include a reference number on the attachment if available. If the attachment consists of several pages, each page should be numbered in Page X of Y format if not electronically done so. And remember, handwritten entries must be accompanied by a signature and date. Always use staples to attach documents together. Glue or paper clips are not acceptable.

Management of Documents through Their Lifecycle

GDP applies to all the different stages of a document's lifecycle. These stages include creation, review, approval, issuing, completion of records, revision, updating, retirement and storage. Storage a.k.a. retention is an important stage and often a legal requirement for medical devices and pharmaceutical products. For consumer OTC medicines, a 5-year retention of quality records often suffices. For implants such as TKRs or Total Knee Replacements, a 90-year retention period is required. This ensures that traceability and a quality record is available if the need arises.

Test Results

This section provides an overview on the correct handling of test results. Test results can be generated from various types of product testing such as visual inspection, dimensional inspection and chemical analysis. The recording of all test results should be completed on an approved form. This is to ensure that the correct information is being recorded and the same approach is taken by different people who might have to complete testing.

Calculations

There are different ways calculations can be completed. Many simple calculations can be done by an individual using a calculator, alternatively, a software package such as Minitab or an Excel sheet can be used to complete complex calculations. It should be clear to the reader what calculation is required, what the formula is and how the calculation is completed.

If the formula used is not included on the sheet, it should be referenced in a controlled document. Care is also required when recording numbers of several decimal places in length, as rounding error can be introduced.

Units of Measurement

The most important thing to remember is consistency in units of measurement when recording data or making calculations. Consult your company procedure if available to determine the correct units of measurement. Many U.S. companies use imperial units e.g. inches, pounds etc. In Europe the International System of Units or SI is used, e.g. millimetres and kilograms.

Batch Records

Batch records document critical information relating to the manufacture of products. Depending on the product, it can include dispensed weights of raw materials. It may also include critical parameters, times and dates of critical steps, in-process test results and so on.

Batch records should be reviewed and checked for:

> ➤ Accuracy
> ➤ Legibility
> ➤ Correct document version
> ➤ Completeness
> ➤ Correct references to supporting documents
> ➤ a unique batch or identification number
> ➤ be dated
> ➤ signed when issued/approved

With reference to ICH Q7, the following requirements are specified in a clear and concise format beneficial to the manufacturer.

"Documentation of completion of each significant step in the batch production records (batch production and control records) should include:
− dates and, when appropriate, times;

− Identity of major equipment (e.g., reactors, driers, mills, etc.) used;

− Specific identification of each batch, including weights, measures, and batch numbers of raw materials, intermediates, or any reprocessed materials used during manufacturing; − actual results recorded for critical process parameters;

— Any sampling performed;

— Signatures of the persons performing and directly supervising or checking each critical step in the operation;

— In-process and laboratory test results;

— Actual yield at appropriate phases or times;

— Description of packaging and label for intermediate or API;

— Representative label of API or intermediate if made commercially available;

— Any deviation noted, its evaluation, investigation conducted (if appropriate) or reference to that investigation if stored separately;

— Results of release testing."

Laboratory Records

Laboratory control records should include complete data derived from all tests conducted to ensure compliance with established specifications and standards, including examinations and assays.

Control records must also provide a description of samples received for testing, including the material name or source, batch number or other distinctive code, date sample was taken, and, where appropriate, the quantity and date the sample was received for testing.

ICH Q7 states the following requirements:

— A statement of or reference to each test method used;

— A statement of the weight or measure of sample used for each test as described by the method; data on or cross-reference to the preparation and testing of reference standards, reagents and standard solutions;

— *A complete record of all raw data generated during each test, in addition to graphs, charts, and spectra from laboratory instrumentation, properly identified to show the specific material and batch tested;*

— *A record of all calculations performed in connection with the test, including, for example, units of measure, conversion factors, and equivalency factors;*

— *A statement of the test results and how they compare with established acceptance criteria;* —

The signature of the person who performed each test and the date(s) the tests were performed; and

— *The date and signature of a second person showing that the original records have been reviewed for accuracy, completeness, and compliance with established standards.*

Complete records should also be maintained for:

— *Any modifications to an established analytical method;*

— *Periodic calibration of laboratory instruments, apparatus, gauges, and recording devices;*

— *All stability testing performed on APIs; and*

— *Out-of-specification (OOS) investigations.*

(Ref: ICH, Q7.)

Materials Management

The key theme of effective materials management is control of materials from the incoming stage through the manufacturing process. Specifications and testing support the control of materials ensuring they are meeting the key quality requirements to allow consistent manufacturing and quality products.

Starting Materials

The designated name of the product and the internal code reference where applicable;

> ➤ Manufacturers batch number
> ➤ the status of the contents (e.g. quarantined, on test, released)
> ➤ the expiry date or a date beyond which retesting is necessary

Packaging Materials

The purchase, handling and control of primary and printed packaging materials should be as for starting materials. Particular attention should be paid to printed packaging materials. They should be stored in secure conditions so as to exclude the possibility of unauthorised access. Roll feed labels should be used wherever possible. Cut labels and other loose printed materials should be stored and transported in separate closed containers so as to avoid mix ups. Packaging materials should be issued for use only by designated personnel following an approved and documented procedure.

Intermediate

Intermediate products can be simply described as raw materials that may have been mixed and processed to some degree or other. Intermediate and bulk products should be kept under appropriate conditions and must be used within specified dates and according to specifications.

Finished Product

Finished products should be held in quarantine until their final release, after which they should be stored as usable stock under conditions established by the manufacturer.

For the approval and maintenance of suppliers of active substances and excipients, the following is required:

Active substances

Supply chain traceability should be established and the associated risks, from active substance starting materials to the finished medicinal product, should be formally assessed and periodically verified. Appropriate measures should be put in place to reduce risks to the quality of the active substance.

The supply chain and traceability records for each active substance (including active substance starting materials) should be available and be retained by the EEA based manufacturer or importer of the medicinal product.

Audits should be carried out at the manufacturers and distributors of active substances to confirm that they comply with the relevant good manufacturing practice and good distribution practice requirements. The holder of the manufacturing authorisation shall verify such compliance either by himself or through an entity acting on his behalf under a contract. For veterinary medicinal products, audits should be conducted based on risk.

Further audits should be undertaken at intervals defined by the quality risk management process to ensure the maintenance of standards and continued use of the approved supply chain.

Excipients

Excipients and excipient suppliers should be controlled appropriately based on the results of a formalised quality risk assessment in accordance with the European Commission 'Guidelines on the Formalised Risk Assessment for Ascertaining the Appropriate Good Manufacturing Practice for Excipients of Medicinal Products for Human Use'.

For each delivery of starting material the containers should be checked for integrity of package, including tamper evident seal where relevant, and for correspondence between the delivery note, the purchase order, the supplier's labels and approved manufacturer and supplier information maintained by the medicinal product manufacturer. The receiving checks on each delivery should be documented.

Prevention of Cross-Contamination

Cross-contamination should be prevented by attention to design of the premises and equipment. This should be supported by attention to process design and implementation of any relevant technical or organisational measures, including effective and reproducible cleaning processes to control risk of cross-contamination.

A Quality Risk Management process, which includes a potency and toxicological evaluation, should be used to assess and control the cross-contamination risks presented by the products manufactured. Factors including; facility/equipment design and use, personnel and material flow, microbiological controls, physico-chemical characteristics of the active substance, process characteristics, cleaning processes and analytical capabilities relative to the relevant limits established from the evaluation of the products should also be taken into account. The outcome of the Quality Risk Management process should be the basis for determining the necessity for and extent to which premises and equipment should be dedicated to a particular product or product family. This may include dedicating specific product contact parts or dedication of the entire manufacturing facility.

Suggested Technical Measures

> ➤ Dedicated manufacturing facility (premises and equipment)
> ➤ Self-contained production areas having separate processing equipment and separate heating, ventilation and air-conditioning (HVAC) systems. It

may also be desirable to isolate certain utilities from those used in other areas

➢ Design of manufacturing process, premises and equipment to minimise opportunities for cross-contamination during processing, maintenance and cleaning

➢ Use of "closed systems" for processing and material/product transfer between equipment

➢ Use of physical barrier systems, including isolators, as containment measures

➢ Controlled removal of dust close to source of the contaminant e.g. through localised extraction

➢ Dedication of equipment, dedication of product contact parts or dedication of selected parts which are harder to clean (e.g. filters), dedication of maintenance tools

➢ Use of single-use disposable technologies

➢ Use of equipment designed for ease of cleaning

➢ Appropriate use of air-locks and pressure cascade to confine potential airborne contaminant within a specified area

➢ Minimising the risk of contamination caused by recirculation or re-entry of untreated or insufficiently treated air

➢ Use of automatic clean in place systems of validated effectiveness

➢ Common general wash areas, separation of equipment washing, drying and storage areas

Suggested Organisational Measures

> Dedicating the whole manufacturing facility or a self-contained production area on a campaign basis (dedicated by separation in time) followed by a cleaning process of validated effectiveness

> Keeping specific protective clothing inside areas where products with high risk of cross-contamination are processed

> Cleaning verification after each product campaign should be considered as a detectability tool to support effectiveness of the Quality Risk Management approach for products deemed to present higher risk

> Depending on the contamination risk, verification of cleaning of non-product contact surfaces and monitoring of air within the manufacturing area and/or adjoining areas in order to demonstrate effectiveness of control measures against airborne contamination or contamination by mechanical transfer

> Specific measures for waste handling, contaminated rinsing water and soiled gowning

> Recording of spills, accidental events or deviations from procedures

> Design of cleaning processes for premises and equipment such that the cleaning processes in themselves do not present a cross-contamination risk

> Design of detailed records for cleaning processes to ensure completion of cleaning in accordance with approved procedures and use of cleaning status labels on equipment and manufacturing areas

> Use of common general wash areas on a campaign basis

Rejection and Re-Use of Materials

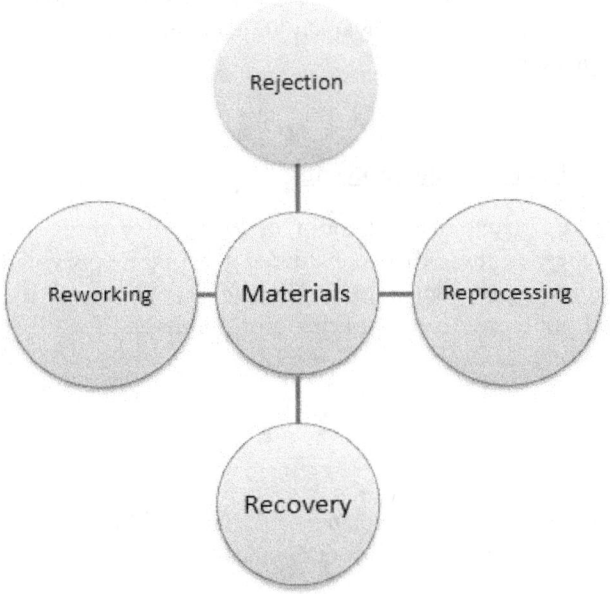

Figure: Material uses

Rejection

Intermediates and components failing to meet established specifications should be identified as such and quarantined according to a procedure. These items can be reprocessed or reworked as described below.

Reprocessing

Reprocessing by repeating a manufacturing step or a chemical or physical process of an established manufacturing process is generally considered acceptable.

Reprocessing should involve evaluation to ensure the quality of product and must not adversely impact the safety of the finished product.

Recovery of Materials and Solvents

Recovery (e.g. from mother liquor or filtrates) of reactants, intermediates, or the API is considered acceptable, provided that approved procedures exist for the recovery and the recovered materials meet specifications suitable for their intended use.

Returns

Records of returned intermediates or APIs should be maintained. For each return, documentation should include:

- Name and address of the consignee

- Intermediate or API, batch number, and quantity returned

- Reason for return

- Use or disposal of the returned intermediate or API

Testing of Materials

The tests performed should be recorded adequately. EU GMP V4 Part 1 Chapter 6: Quality Control recommends the following information as a minimum:

- ➤ Name of the material or product and, where applicable, dosage form
- ➤ Batch number and, where appropriate, the manufacturer and/or supplier
- ➤ References to the relevant specifications and testing procedures
- ➤ Test results, including observations and calculations, and reference to any certificates of analysis
- ➤ Dates of testing
- ➤ Initials of the persons who performed the testing
- ➤ Initials of the persons who verified the testing and the calculations, where appropriate

Sampling Checklist

The sample taking should be done and recorded in accordance with approved written procedures that describe:

- ➤ The method of sampling
- ➤ The equipment to be used
- ➤ The amount of the sample to be taken
- ➤ Instructions for any required sub-division of the sample
- ➤ The type and condition of the sample container to be used

- ➢ The identification of containers sampled
- ➢ Any special precautions to be observed, especially with regard to the sampling of sterile or noxious materials
- ➢ The storage conditions
- ➢ Instructions for the cleaning and storage of sampling equipment

Validation

The term 'validation lifecycle' refers to the entire lifecycle, beginning with the initial requirements of a product or process and identifying CPPs and CQAs. The cycle continues through Commissioning and Qualification (C&Q), PQ and PV, requalification and ends with decommissioning or the end of life of a product line.

Process Validation is defined as "establishing documented evidence which provides a high degree of assurance that a specific process consistently produces a product meeting its predetermined specifications and quality attributes."

Why Is Validation Required?

There are several factors that require validation activities within the life science industry. Above all, validation works to ensure patient safety and products that are fit for purpose and reliable time after time. However, regulations provide the legal incentive to validate processes within the medical device and medicinal industry. Validation activity also has a secondary effect of fostering consistency in introducing new products and processes across different departments and sites.

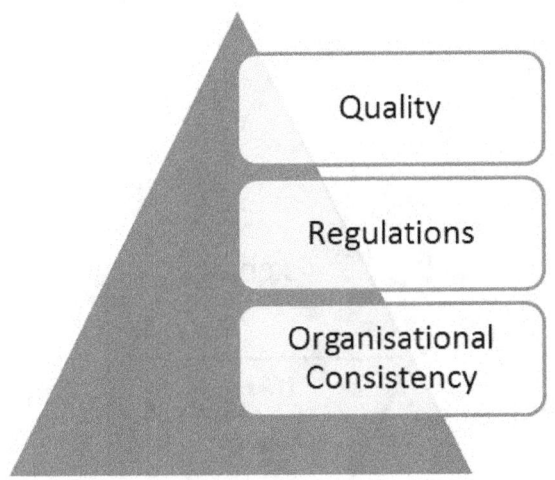

Figure: Drivers of Validation

In many respects, regulations are the key driver, and as with any legal requirement, the manufacturer is required to fulfil their statutory obligations. Notified bodies such as the UK, MHRA, US FDA specify rules and guidance in respect of pharmaceutical and medical products.

As the largest economy is the world, the United States is a leading user of regulated products. In turn, many countries throughout the world manufacture products with the intent to supply the US market. Therefore, manufacturers must meet the regulatory requirements set down by the US FDA. For medical devices, 21 CFR Part 820 requires validation to be completed for equipment and processes. For pharmaceuticals, 21 CFR Part 211 also makes provision for validation. In Europe, validation for the manufacture of medicinal products is a requirement of EU GMP V4, Medicinal Products for Human and Veterinary Use.

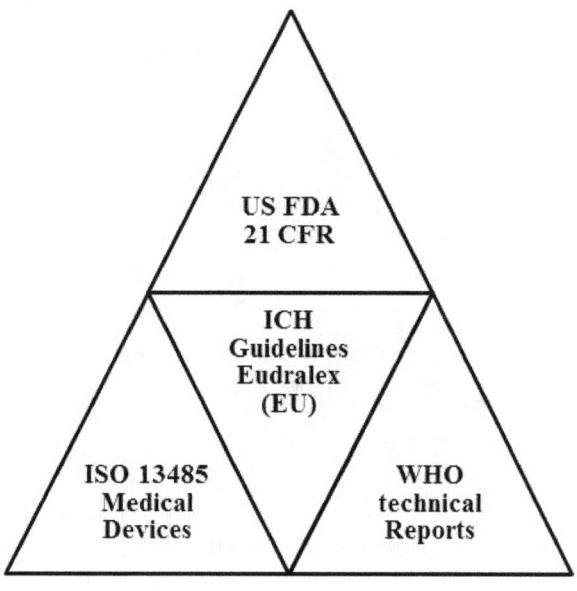

Figure: Key regulations and standards

<u>PICS/s</u>

The PIC/s guidance on validation is limited to describing the basic requirements of a validation programme such as having a validation procedure, qualifying a new process so that it is "suitable and "consistent".

PE 009-13 (Part I), Chapter 5, states:

➢ *"Validation studies should reinforce Good Manufacturing Practice and be conducted in accordance with defined procedures. Results and conclusions should be recorded.*

➢ *When any new manufacturing formula or method of preparation is adopted, steps should be taken to demonstrate its suitability for routine processing. The defined process, using the materials and equipment specified, should be shown to yield a product consistently of the required quality.*

➢ *Significant amendments to the manufacturing process, including any change in equipment or materials, which may affect product quality and/or the reproducibility of the process should be validated.*

➢ *Processes and procedures should undergo periodic critical revalidation to ensure that they remain capable of achieving the intended results."*

EudraLex

"Validation studies should reinforce Good Manufacturing Practice and be conducted in accordance with defined procedures. Results and conclusions should be recorded.

When any new manufacturing formula or method of preparation is adopted, steps should be taken to demonstrate its suitability for routine processing. The defined process, using the materials and equipment specified, should be shown to yield a product consistently of the required quality.

Significant amendments to the manufacturing process, including any change in equipment or materials, which may affect product quality and/or the reproducibility of the process, should be validated.

Processes and procedures should undergo periodic critical re-validation to ensure that they remain capable of achieving the intended results." (Ref: GMP V4 Part I, Chapter 5)

FDA

For medical devices, FDA 21 CFR Subpart G, Part 820 specifies the requirements for full verification or validation:

(a) Where the results of a process cannot be fully verified by subsequent inspection and test, the process shall be validated with a high degree of assurance and approved according to established procedures. The validation activities and results, including the date and signature of the individual(s) approving the validation and where appropriate the major equipment validated, shall be documented.

(b) Each manufacturer shall establish and maintain procedures for monitoring and control of process parameters for validated processes to ensure that the specified requirements continue to be met.

(1) Each manufacturer shall ensure that validated processes are performed by qualified individual(s).
(2) For validated processes, the monitoring and control methods and data, the date performed, and, where appropriate, the individual(s) performing the process or the major equipment used shall be documented.

(c) When changes or process deviations occur, the manufacturer shall review and evaluate the process and perform revalidation where appropriate. These activities shall be documented.

For medicinal/ pharmaceutical drug products 21 CFR 211.100(a) and 211.110(a) requires that drug products be produced with a high degree of assurance of meeting all the attributes they are intended to possess.

WHO

WHO GMP guidance requires that, each pharmaceutical company identifies what qualification and validation work is required to prove that the critical aspects of their particular operation are controlled. A validation plan should identify and describe what activities are required to be undertaken.

Annex 2 identifies the key requirements with regard to qualification and validation to ensure:

➢ the premises, supporting utilities, equipment and processes have been designed in accordance with the requirements for GMP (design qualification or DQ)

➢ the premises, supporting utilities and equipment have been built and installed in compliance with their design specifications (installation qualification or IQ)

➢ the premises, supporting utilities and equipment operate in accordance with their design specifications (operational qualification or OQ)

➤ a specific process will consistently produce a product meeting its predetermined specifications and quality attributes (process validation or PV, also called performance qualification or PQ)

ICH

ICH Q7 provides arguably provides the greatest amount of detail with regard to validation in a GMP environment.

Similar to other organisations, ICH requires a company to develop a "validation policy" to describe and document approaches to validations etc. ICH gives good guidance on validation in respect of active pharmaceutical ingredients and the importance of critical process parameters and critical quality attributes. The approaches to process validation (Prospective, Concurrent etc.) also align with FDA requirements.

Key Considerations for Validation of APIs

➤ Defining critical product attributes of APIs
➤ Identifying process parameters that could affect the critical quality attributes of APIs
➤ Process validation (PV should provide documented evidence that the process, operated within established parameters, can perform effectively and reproducibly to produce an intermediate or API meeting its predetermined specifications and quality attributes)

ICH Q7 Key Definitions Regarding Qualification/Validation

"Design Qualification (DQ): documented verification that the proposed design of the facilities, equipment, or systems is suitable for the intended purpose.

Installation Qualification (IQ): documented verification that the equipment or systems, as installed or modified, comply with the approved design, the manufacturer's recommendations and/or user requirements.

Operational Qualification (OQ): documented verification that the equipment or systems, as installed or modified, perform as intended throughout the anticipated operating ranges.

Performance Qualification (PQ): documented verification that the equipment and ancillary systems, as connected together, can perform effectively and reproducibly based on the approved process method and specifications."

The Four Types of Process Validation

Process validation is a regulatory requirement of Good Manufacturing Practices (GMPs) for both pharmaceuticals (21CFR 211) and medical devices (21 CFR 820).

Prospective validation

Establishing documented evidence in advance of process implementation that a process or system operates as intended. This is the preferred approach and is most common when new products must be validated before commercial manufacturing.

Concurrent Validation

Establishing documented evidence that a processes operates as intended, based on information generated during process implementation. Concurrent means that the outputs and performance of the system is monitored at the same time as manufacturing which can include commercial lots.

Retrospective Validation

Retrospective validation is used for facilities or processes that have not completed formal validation. Historical data or a retrospective review can provide the evidence that the process or facility is operated as intended. This type of validation is uncommon.

Revalidation

Revalidation involves the re-execution of validation activities in order to maintain a validated state. This can be a result of substantial changes to product attributes or specification or changes to the manufacturing process itself. Other reasons a partial or full revalidation may be required involve instances where product quality issues have increased.

Stages of Process Validation

Process validation can be divided into in three stages:

Stage 1 – Process Design: The commercial manufacturing process is defined during this stage based on knowledge gained through development and scale-up activities.

<u>Stage 2</u> – Process Qualification: During this stage, the process design is evaluated to determine if the process is capable of reproducible commercial manufacturing.

<u>Stage 3</u> – Continued Process Verification: Ongoing assurance is gained during routine production that the process remains in a state of control.

Before any batch from the process is commercially distributed for use by consumers, a manufacturer should have gained a high degree of assurance in the performance of the manufacturing process such that it will consistently produce APIs and drug products meeting those attributes relating to identity, strength, quality, purity, and potency. The assurance should be obtained from objective information and data from laboratory; pilot and/or commercial scale studies. Information and data should demonstrate that the commercial manufacturing process is capable of consistently producing acceptable quality products within commercial manufacturing conditions. A successful validation programme depends upon information and knowledge from product and process development. This knowledge and understanding is the basis for establishing an approach to control of the manufacturing process that results in products with the desired quality attributes. Understanding variation and knowing how to detect and control it is therefore a key element to maintaining robust processes and systems.

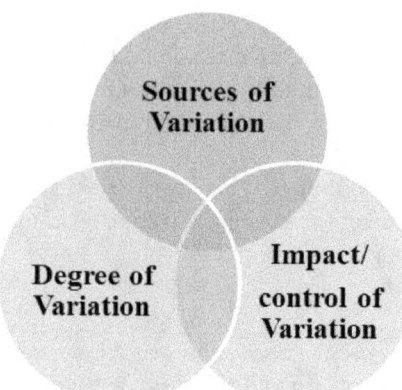

Figure: Variation

Manufacturers should understand the sources of variation, the degree of variation, the impact of variation and how to detect and control variation. Each manufacturer should judge whether it has gained sufficient understanding to provide a high degree of assurance in its manufacturing process to justify commercial distribution. Focusing exclusively on qualification efforts without also understanding the manufacturing process and associated variations may not lead to adequate assurance of quality. After establishing and confirming the process, manufacturers must maintain the process in a state of control over the life of the process, even as materials, equipment, production environment, personnel, and manufacturing procedures change. Manufacturers should use ongoing programmes to collect and analyse product and process data to evaluate the state of control of the process. These programs may identify process or product problems or opportunities for process improvements that can be evaluated and implemented through some of the activities described in Stages 1 and 2.

Manufacturers of legacy products can take advantage of the knowledge gained from the original process development and qualification work as well as manufacturing experience to continually improve their processes. Implementation of the recommendations in this guidance for legacy products and processes would likely begin with the activities described in Stage 3.

Stage 1 — Process Design: Process design is the activity of defining the commercial manufacturing process that will be reflected in planned master production and control records.

The goal of this stage is to design a process suitable for routine commercial manufacturing that can consistently deliver a product that meets its quality attributes by building and capturing process knowledge and understanding. Generally, early process design experiments do not need to be performed under the CGMP conditions required for drugs intended for commercial manufacturing and supply.

Stage 2 (process qualification) and Stage 3 (continued process verification should be conducted in accordance with sound scientific methods and principles, including good documentation practices. Decisions and justification of the controls should be sufficiently documented and internally reviewed to verify and preserve their value for use or adaptation later in the lifecycle of the process and product. Although often performed at small-scale laboratories, most viral inactivation and impurity clearance studies cannot be considered early process design experiments.

Viral and impurity clearance studies intended to evaluate and estimate product quality at commercial scale should have a level of quality unit oversight that will ensure that the studies follow sound scientific methods and principles and the conclusions are supported by the data.

Product development activities provide key inputs to the process design stage, such as the intended dosage form, the quality attributes, and a general manufacturing pathway. Process information available from product development activities can be leveraged in the process design stage. The functionality and limitations of commercial manufacturing equipment should be considered in the process design, as well as predicted contributions to variability posed by different component lots, production operators, environmental conditions, and measurement systems in the production setting. Design of Experiment (DOE) studies can help develop process knowledge by revealing relationships, including multivariate interactions, between the variable inputs (e.g., component characteristics or process parameters) and the resulting outputs (e.g. in-process material, intermediates, or the final product).

Risk analysis tools can be used to screen potential variables for DOE studies to minimise the total number of experiments conducted while maximising knowledge gained.

These activities also provide information that can be used to model or simulate the commercial process. Computer-based or virtual simulations of certain unit operations or dynamics can provide process understanding and help avoid problems at commercial scale. It is important to understand the degree to which models represent the commercial process, including any differences that might exist, as this may have an impact on the relevance of information derived from the models. It is essential that activities and studies resulting in process understanding be documented. Establishing a strategy for process control process knowledge and understanding is the basis for establishing an approach to process control for each unit operation and the process overall.

Strategies for process control can be designed to reduce input variation, adjust for input variation during manufacturing (and so reduce its impact on the output), or combine both approaches. Process controls address variability to ensure quality of the product. Controls can consist of material analysis and equipment monitoring at significant processing. Decisions regarding the type and extent of process controls can be aided by earlier risk assessments, then enhanced and improved as process experience is gained. The planned commercial production and control records, which contain the operational limits and overall strategy for process control, should be carried forward to the next stage for confirmation.

Stage 2 — Process Qualification: During the process qualification (PQ) stage of process validation, the process design is evaluated to determine if it is capable of reproducible commercial manufacture.

This stage has two elements: (1) design of the facility and qualification of the equipment and utilities and (2) process performance qualification (PPQ). During Stage 2, CGMP-compliant procedures must be followed. Successful completion of Stage 2 is necessary before commercial distribution. Products manufactured during this stage, if acceptable, can be released for distribution.

Design of a facility, qualification of utilities and equipment and proper design of a manufacturing facility are required under part 211, subpart C, of the CGMP regulations on buildings and facilities. It is essential that activities performed to ensure proper facility design and commissioning precede PPQ.

Here, the term qualification refers to activities undertaken to demonstrate that utilities and equipment are suitable for their intended use and perform properly. These activities necessarily precede manufacturing products at the commercial scale. Qualification of utilities and equipment generally includes the following activities:

➤ Selecting utilities and equipment construction materials, operating principles, and performance characteristics based on whether they are appropriate for their specific uses.

➤ Verifying that utility systems and equipment are built and installed in compliance with the design specifications (e.g. built as designed with proper materials, capacity, and functions, and properly connected and calibrated).

> Verifying that utility systems and equipment operate in accordance with the process requirements in all anticipated operating ranges. This should include challenging the equipment or system functions while under load comparable to that expected during normal operation.

It should also include the performance of interventions, stoppage, and start-up as is expected during routine production. Operating ranges should be shown capable of being held as long as would be necessary during routine production. Qualification of utilities and equipment can be covered under individual plans or as part of an overall project plan.

The plan should consider the requirements of use and can incorporate risk management to prioritise certain activities and identify a level of effort in both the performance and documentation of qualification activities. Planning must also give consideration to the following points:

Design of facilities and the qualification of utilities and equipment, personnel training and qualification, and verification of material sources (components and container/closures), if not previously accomplished.

Review and approval of the protocol by appropriate departments and the quality unit. PPQ protocol execution and report xxecution of the PPQ protocol should not begin until the protocol has been reviewed and approved by all appropriate departments, including the quality unit. Any departures from the protocol must be made according to established procedure or provisions in the protocol. Such departures must be justified and approved by all appropriate departments and the quality unit before implementation (§ 211.100).

The commercial manufacturing process and routine procedures must be followed during PPQ protocol execution (§§ 211.100(b) and 211.110(a)). The PPQ lots should be manufactured under normal conditions by the personnel routinely expected to perform each step of each unit operation in the process. Normal operating conditions should include the utility systems (e.g., air handling and water purification), material, personnel, environment, and manufacturing procedures. A report documenting and assessing adherence to the written PPQ protocol should be prepared in a timely manner after the completion of the protocol.

This report should:

> Discuss and cross-reference all aspects of the protocol.
> Summariae data collected and analyse the data, as specified by the protocol.
> Evaluate any unexpected observations and additional data not specified in the protocol.

➢ Summarise and discuss all manufacturing non-conformances such as deviations, aberrant test results, or other information that has bearing on the validity of the process.

➢ Describe in sufficient detail any corrective actions or changes that should be made to existing procedures and controls.

➢ State a clear conclusion as to whether the data indicates the process met the conditions established in the protocol and whether the process is considered to be in a state of control. If not, the report should state what should be accomplished before such a conclusion can be reached. This conclusion should be based on a documented justification for the approval of the process, and release of lots produced by it to the market in consideration of the entire compilation of knowledge and information gained from the design stage through the process qualification stage.

➢ Include all appropriate department and quality unit review and approvals.

Stage 3 — Continued Process Verification

The goal of the third validation stage is continual assurance that the process remains in a state of control (the validated state) during commercial manufacture. A system or systems for detecting unplanned departures from the process as designed are essential to accomplish this goal. Adherence to the CGMP requirements, specifically, the collection and evaluation of information and data about the performance of the process, will allow detection of undesired process variability.

Evaluating the performance of the process identifies problems and determines whether action must be taken to correct, anticipate, and prevent problems so that the process remains in control (§ 211.180(e)). An ongoing programme to collect and analyse product and process data that relate to product quality must be established (§ 211.180(e)).

The data collected should include relevant process trends and quality of incoming materials or components, in-process material, and finished products. The data should be statistically trended and reviewed by trained personnel. The information collected should verify that the quality attributes are being appropriately controlled throughout the process. It is recommend that a statistician or person with adequate training in statistical process control techniques develop the data collection plan and statistical methods and procedures used in measuring and evaluating process stability and process capability.

Procedures should describe some references that may be useful including the following:

> ASTM E2281-03 "Standard Practice for Process and Measurement Capability Indices,"

> ASTM E2500-07 "Standard Guide for Specification, Design, and Verification of Pharmaceutical and Biopharmaceutical Manufacturing Systems and Equipment."

> ASTM E2709-09 "Standard Practice for Demonstrating Capability to Comply with a Lot Acceptance Procedure."

Production data should be collected to evaluate process stability and capability. The quality unit should review this information. If properly carried out, these efforts can identify variability in the process and/or signal potential process improvements. Good process design and development should anticipate significant sources of variability and establish appropriate detection, control, and/or mitigation strategies, as well as appropriate alert and action limits. However, a process is likely to encounter sources of variation that were not previously detected or to which the process was not previously exposed. Many tools and techniques, some statistical and others more qualitative, can be used to detect variation, characterise it, and determine the root cause.

Leading manufacturers should use quantitative, statistical methods whenever appropriate and feasible. Scrutiny of intra-batch as well as inter-batch variation is part of a comprehensive continued process verification programme under § 211.180(e).

Best practices ensures continued monitoring and sampling of process parameters and quality attributes at the level established during the process qualification stage until sufficient data are available to generate significant variability estimates. These estimates can provide the basis for establishing levels and frequency of routine sampling and monitoring for the particular product and process.

Monitoring can then be adjusted to a statistically appropriate and representative level. Process variability should be periodically assessed and monitoring adjusted accordingly. Variation can also be detected by the timely assessment of defect complaints, out-of-specification findings, process deviation reports, process yield variations, batch records, incoming raw material records and adverse event reports.

Production line operators and quality unit staff should be encouraged to provide feedback on process performance. We recommend that the quality unit meet periodically with production staff to evaluate data, discuss possible trends or undesirable process variation, and coordinate any correction or follow-up actions by production.

Data gathered during this stage might suggest ways to improve and/or optimise the process by altering some aspect of the process or product, such as the operating conditions (ranges and set-points), process controls, component, or in-process material characteristics. A description of the planned change, a well-justified rationale for the change, an implementation plan, and quality unit approval before implementation must be documented (§ 211.100). Depending on how the proposed change might affect product quality, additional process design and process qualification activities could be warranted.

Maintenance of the facility, utilities, and equipment is another important aspect of ensuring that a process remains in control. Once established, qualification status must be maintained through routine monitoring, maintenance, and calibration procedures and schedules (21 CFR part 211, certain manufacturing changes may call for formal notification to the agency before implementation, as directed by existing regulations (see, e.g., 21 CFR 314.70 and 601.12).

The equipment and facility qualification data should be assessed periodically to determine whether re-qualification should be performed and the extent of that re-qualification. Maintenance and calibration frequency should be adjusted based on feedback from these activities.

Equipment Validation

Validation is a legal and regulatory requirement for the manufacture of medicinal products. The area of validation can be sub-divided into two elements. Equipment Qualification (EQ) and Process Validation. Equipment qualification ensures that equipment operates as intended and is installed in accordance with the manufacturer's recommendation. Process Validation involves the provision of documented evidence to confirm a particular process performs consistently and meets pre-determined specifications.

All equipment that can impact the quality of product is subject to validation, hence equipment and systems used in aseptic manufacturing must undergo equipment and process validation.

Installation Qualification (IQ) protocols should include verification that all utilities are installed correctly to the manufacturer's recommendations. All sitting and mechanical connections should also be confirmed as adequate. Other key tests and verifications include:

- ➢ Documentation of Materials of Construction (MOC)
- ➢ Calibration of equipment based instrumentation
- ➢ Spare parts listing
- ➢ Preventative maintenance schedule creation
- ➢ Electrical installation verification
- ➢ Health and safety assessment
- ➢ Ergonomic assessment
- ➢ Documenting software and hardware
- ➢ Backup of software
- ➢ Backup of recipes (sterilisation, bio-decontamination etc.)

The system User Requirements Specification (URS) should provide the basis of testing and must be fulfilled during the course of validation.

The ultimate goal of Equipment Qualification is to ensure that equipment is fit for its intended use. Therefore, equipment is validated to confirm it functions as intended and meets all requirements to manufacture product safely and consistently. FDA requires that "Each manufacturer shall ensure that all equipment used in the manufacturing process meets specified requirements and is appropriately designed, constructed, placed and installed to facilitate maintenance, adjustment, cleaning and use." In other words all manufacturing equipment, support facilities, measuring and test equipment shall be "qualified". (FDA 21 CFR 820.70 (g))

Equipment qualification protocols are developed to document this testing and hence provide evidence on the functionality and consistency of the equipment. There are two distinct parts within the scope of equipment qualification, installation qualification and operational qualification. Often these subparts are abbreviated to IQ and OQ. Other combinations such as IOQE and IQ/OQ can be encountered within industry. This is often defined in a company's procedure or SOP relating to equipment validation.

A User Requirements Specification (URS) is often used to document the "specified requirements" of a particular piece of equipment. A URS can then be used as an input document when equipment qualification is required. While a URS document can be extensive covering areas such as equipment functionality, utility requirements, safety features, software specs etc. not all requirements documented in a URS will need to be verified or validated. Critical requirements should be identified early and should always be verified.

In short, equipment qualification is confirmation via documented evidence that the particular requirements for a specific intended use can be consistently fulfilled under anticipated conditions.

Often referred to as the three "Cs" of validation; confirmation, consistency and conditions (anticipated) are key themes that validation must address. Confirmation is addressed by the process of completing a formal validation. When it's done, it is documented and available for review to auditors. To assess consistency, there must be a number of batches or "runs". Typically, there are minor batch-to-batch differences or variations between batches. These differences can be as a result of setup or raw material differences. Process validation must ensure that despite minor changes, there is consistency between batches, with product meeting specifications. Controlled or anticipated conditions are the machine or process settings that are known, documented and controlled during the manufacture of products.

Materials of Construction (MOC)

The materials of construction and evidence of the same (certificates) form part of installation also. Materials must be fit for the intended purpose and compatible with products and manufacturing agents that come into contact with them.

For example, fermenters are made of materials that are suited to the use of steam sterilisation techniques and regular cleaning. Such materials can be classed as both non-reactive and non-absorptive surfaces. Most aseptic processing equipment that incorporates product contact surfaces is made of high grade stainless steel. Cheaper classifications of stainless steel can be used for jacketing and other non-product contact areas.

All interior product contact surfaces should be polished to a "mirror" finish. Welds also need to be finished in a similar manner. Electro polishing provides a better quality surface finish than mechanical polishing.

As with any chemical reaction, factors such as temperature, pH and oxygen concentration can impact the performance and yield. To ensure the optimum conditions are maintained, it is important to monitor and control such parameters and factors. By far the most common these days is automatic control of systems and equipment with automatic feedback and adjustment.

Operational Qualification (OQ) is the second component of equipment qualification. This is "Establishing by documented evidence that the equipment operates per specifications and over the required ranges and to required tolerances". Equipment is also tested to ensure alarms and controls operate as required and intended. Some typical checks included in an equipment-operational qualification are testing of alarms, control system testing, utility failures and functional and operational testing.

Suggested IQ/OQ Verifications/Tests

Standing Operating Procedures (SOPs)

SOPs are designed to provide formal documented instruction on how to execute tasks or operate equipment and machinery. While each company will require different headings, a work instruction or SOPs should cover set-up, system operation, cleaning and shutdown to name but a few.

Test Instrumentation Calibration

External test devices such as temperature probes, volt meters, lux meters and particle counters may be required to take measurements during an equipment qualification. Test instrumentation should have a suitable range, resolution and accuracy. Certificates of calibration should also be available, with calibration conforming to a recognised external standard. Information such as the serial number, model number and manufacturer should be recorded for reference and traceability.

Equipment Based Instrumentation Calibration

All equipment based instruments must be calibrated as part of equipment validation or in advance of it. Instruments should have unique calibration ID.

Electrical Checks

Appropriate connections and earthing checks are required. A review of electrical drawings to ensure the physical status is as per drawings and specifications is also necessary. Cables and electrical hazards should also be appropriately labelled.

Mechanical Checks

Ensure the systems are fixed, fastened and integrated mechanically. Safety guards and barriers should also be in place where required.

Pneumatic Checks

Verify the proper supply and integration of compressed air. Supply should be leak-free, and regulated with filters and water traps fitted as required.

Documentation

Documentation requires verification that design, operation and maintenance documents have been received from the manufacturer and are stored appropriately.

Ergonomics

Controls and HMIs should be positioned to facilitate ease of use and should be identified clearly.

Health and Safety

Hazards are identified and guarded, pin points are identified. No trip hazards are evident and emergency stops function.

Software

Equipment installation software checks should record the names and version numbers of all software. HMI software, PLC software, application software. Provision should be made for disaster recovery and backup.

Hardware

Computer hardware should be recorded to include the model, manufacturer, serial number and specification details.

Environmental

Any features detailed in the URS relating to environmental requirements need to be verified during IQ/OQ. For example, automatic shutdown after periods of inactivity. Heating and cooling systems should also be appropriately insulated.

Alarms

Automated processes such as sterilisation tunnels, Autoclaves, filling machines and isolators typically have many alarms and controls. Alarms can be categorised as critical or non-critical to the process or product. Depending on the vendor or manufacturer alarms can also be grouped according to the type of alarm (EHS, process, mechanical, pneumatic and so on). Alarms should be tested to ensure the right action by the machine is taken, the process comes to a safe stop, and that the alarm can be acknowledged and the alarm condition cleared.

Utility Failure

Also referred to as provoke testing, utility failure of compressed air, fume extraction, electrical supply etc. is designed to ensure that in the event of failure during commercial manufacturing, the equipment comes to a safe stop and can be brought back into use upon recovery of the utility.

Fixtures

Materials of construction must be suitable for the intended use. In aseptic processing, high grades of stainless steel (316L) are the preferred material of use.

Functional Tests

The individual functions of equipment must be verified during commissioning and qualification.

Suggested Prerequisites to Equipment Qualification

Prior to formal Equipment Installation and Operational Qualification (IQOQ), there are a number of engineering activities that can be completed. Although work is required up front in order to complete these activities such as preparing engineering test protocols, it will benefit the qualification stage. The completion of some or all of the above activities will help identify issues prior to formal equipment qualification. Essentially, an FAT Protocol is like an early draft of an IQOQ-Equipment Protocol.

Factory Acceptance Testing (FAT)

FAT or Factory Acceptance Test is an engineering activity. The purpose of the FAT is to verify that the equipment or system meets the requirements of the URS. From the validation engineer's perspective, it can be a learning activity and an opportunity to gather data, documentation and supporting design documents that will prove valuable during the equipment and process validation of the equipment. SAT is an engineering activity that is completed at the site of the vendor or equipment manufacturer, post FAT.

Site Acceptance Testing (SAT)

Site acceptance testing is also an engineering activity conducted when equipment arrives onsite. Depending on the company, SAT can be completed by the vendor or by the purchasing company. It consists of a series of installation and operational checks to ensure the equipment has not suffered any damage or deterioration between the disassembly, crating, shipping and delivery of the equipment.

Equipment Qualification (EQ) Protocols

Protocol Preparation
Thorough and careful preparation is critical in order to successfully complete a qualification without deviations. In the preparation of the protocol, the URS is a key document. Often quotations, design documents, vendor drawings and owner manuals can contribute to the test and verifications to be completed during EQ.

Protocol Approval

Approval is always required prior to executing an equipment qualification protocol. Approvers should be aware that they are signing for the accuracy and content of the whole document. It is strongly advised that prior to final approval and execution of a protocol, a dry-run or trial is completed to ensure the test methods and acceptance criteria are accurate.

Post-Execution Review

Upon execution of a protocol, timely review is advised in order to catch any errors or omissions. The person reviewing the protocol should not be the same person who performed the test. This review is best completed by a quality engineer; however, each organisation should identify personnel responsible for EQ reviews.

Some points to remember when completing protocols:

➢ Ensure the protocol is fully approved prior to execution
➢ Ensure personnel are trained on the protocol (if required) and trained to the specific work instructions
➢ Ensure all team members, contractors etc. sign the signature log
➢ Observe safety precautions and wear PPE as required
➢ Ensure other employees that may be impacted by qualifications are aware that a qualification is in progress
➢ Ensure that any test product required in support of the qualification is identified, segregated and stored to internal standards
➢ Check that accurate work instructions are available (some companies may allow redlined copies to be used)
➢ Complete all tests in the protocol
➢ Always use indelible ink
➢ Carefully check each result against any acceptance criteria

➤ Ensure all test equipment is validated within calibration prior to use

➤ Handwritten comments should be signed and dated per GDP

➤ Deviations should be written up at the time of observation

➤ Data records and attachments should be identified with the protocol number signed, paginated and dated

➤ When data is transcribed it should be verified by a second person. The source of the data should also be recorded

➤ All product manufactured should have relevant batch documentation as per normal production conditions

Equipment Qualification Reports

On completion, Equipment Installation and Operational Protocol reports are required. The format of any report largely depends on company specific procedures. If the protocol is an executable document (results are handwritten in) then the executed version can be deemed the report. A summary report may be required but this depends on the requirements within your company or organisation.

The typical requirements of a completed EQ validation include:

- ➢ Equipment Qualification Protocol
- ➢ Equipment Qualification Protocol (executed)
- ➢ Raw data
- ➢ Attachments (examples of attachments include: material certs, calibration certs, CE cert and MSDS)

Software Validation

Where there is the potential to affect product conformance to requirements or where software or IT systems provide support to aspects of quality management, validation is required.

Most companies categorise software validations to account for the different applications of software and IT systems. For example, enterprise systems, such as the drawing package SolidWorks would be validated in a different manner to manufacturing systems that contain software (a.k.a. embedded software).

Embedded software is where the software is integrated into the manufacturing equipment. Embedded software is typically validated during the equipment qualification stage, process validation stage or test method validation. Enterprise software falls outside of equipment or process validation but does require validation if it impacts product quality or is used to make quality decisions. Standalone systems such as ERP (Enterprise Resource Planning) systems also require validation.

Software Validation and GAMP

Good Automated Manufacturing Practice (GAMP) is a set of guidelines for manufacturers and users of automated systems in regulated industries. Specifically, the medical device, pharmaceutical and biopharmaceutical industries. The application of GAMP and validation of automated systems in manufacturing helps ensure that regulated medical devices and medicinal products have the required quality and are manufactured according to good practices, meet regulatory and legal requirements and ensure patient safety. GAMP ensures quality is in-built into each stage of the manufacturing process. Therefore, GAMP has a place in all aspects of automation and production, including the handling of raw materials, control of facilities and equipment etc.

Key Terms

Automated System: Term used to cover a broad range of systems, including automated manufacturing equipment, control systems, automated laboratory systems manufacturing execution systems and computers running laboratory or manufacturing database systems. The automated system consists of the hardware, software and network components, together with the controlled functions and associated documentation. Automated systems are sometimes referred to as computerised systems; in this guide the two terms are synonymous.

Commercial Off-the-Shelf (COTS) Configurable Programs: Stock programs that can be configured to specific user applications by "filling in the blanks", without (COTS) altering the basic program.

Computer System Validation: A process that confirms by examination and provision of objective evidence that the computer system conforms to user needs and intended uses. System validation is a process for achieving and maintaining compliance with GxP regulations and fitness for intended use by adoption of life-cycle activities, deliverables, and controls.

GAMP 5: A set of guidelines that offers a risk-based approach to ensuring the compliance of GxP impacting computerised systems.

V-Model: A development process which sets out a roadmap of stages and deliverables during a project.

21 CFR Part 820: FDA requirements pertaining to medical devices:

User Requirement Specification, URS: The URS is a critical document that defines the requirements of the computerised system and agreement to the requirements.

Software Requirement Specification, SRS: An SRS can be written to interpret the requirements of a URS and how they relate to the requirement or how the requirement is met in practical terms regarding software.

Functional Design Specification, FDS: A functional design specification is a document that specifies how particular requirements are met – this can be a combination of how the equipment/process operates mechanically/automatically etc. An FDS is typically written in response to a URS.

Computer System Validation Life-Cycle

The computer system validation life cycle refers to all activities from initial concept to retirement of a computer system. The life-cycle of the system includes the defining of and performance of activities in a systematic way from conception, requirements, development or configuration, testing, release and operational use.

The four GAMP life-cycle phases include:

➢ Concept
➢ Planning and project stage
➢ Operation
➢ Retirement

The concept stage is concerned with understanding the need or the problem to be addressed. We will see that the User Requirement Specification (along with other specifications) and the initial risk assessment help to drive a project forward in a systematic manner. The most common life-cycle approach for computerised and automated systems is the V-Model. The GAMP based V-model lays out a roadmap which facilitates the validation of equipment and automated systems.

The planning and project stage involves the planning of the validation effort required to implement the system into the business area(s) based on identification and approval of system concept. This phase includes assessments of the regulatory and system risks, supplier assessment, development of validation strategies, identification of deliverables that will be generated, definition of the business process the system will support as well as the user requirements which the system will fulfil.

Design, development and configuration of the hardware and software is also required to meet the system requirements as per specifications. In case of custom software components, this effort could also include detailed software design and developmental testing to ensure readiness for verification testing.

Verification – This effort confirms that specifications have been met and releases the system for use. This phase will involve multiple stages of reviews and testing depending on the system type, the development method applied and its use. Once verification activities have begun any changes to the system must be captured through change control.

On successful completion of the verification activities, the system is then released for effective use. The test strategy and other verification activities will vary widely between simple equipment and more complex customisable/configurable systems. The verification and validation approach is typically agreed and detailed in at the validation planning stage. The VP can be updated accordingly as the project develops with more detail being added. Alternatively, a test strategy document or matrix could be written to provide more specific test plans.

Verification deliverables vary based on the complexity and level or customisation of the system in question. Corporate or company specific procedures also shape the required activities to be completed and reported. Some generic deliverables are listed below:

> Approval, execution and review of test protocols
> Writing and approving SOPs for operation and maintenance of the system
> Traceability matrix
> Completion of any risk mitigations (e.g. updates to FMEA etc.)
> Validation summary report(s)

Validation reporting requirements varies depending upon the scope of the system and should also be driven by a procedure and template. The validation plan can also outline the deliverables and what needs to be addressed in the report. A Validation Summary Report (VSR) shall be written which summarises the results of executing the VP, the documents created for the validation activities summarises (or points to summaries) of the testing performed. Finally, the VSR indicates the acceptance of the system/equipment by the user, by the project team and states that the equipment is released for commercial operation/ production.

The operation phase supports the need to maintain compliance and fitness for intended use after the system is released for normal use. It is important to ensure the system remains within a continued validated state. All proposed or necessary changes to the system must be assessed and controlled as part of a change control process. Once the system has been accepted and released for use, the operation phase begins. This phase consists of maintaining the system's compliant state and fitness for intended use through the control of the procedures supporting the system's operational use.

During the operation phase the below activities are typically completed:

- ➤ Ongoing training
- ➤ Preventative maintenance
- ➤ Service management and performance monitoring.
- ➤ Change control
- ➤ Periodic review
- ➤ Maintaining system security
- ➤ Records management
- ➤ Calibration

The retirement phase involves the planning and proper management of activities relating to the removal of systems from service (shutdown). The retirement should take into account the storage of any data and any data migration that needs to occur prior to retirement. The retirement plan, if needed, will outline the retirement strategy from the roles and activities that will be conducted to the removal of the system for use. A Retirement Summary Report is produced that documents the results of the activities defined in the retirement plan including:

> ➤ Retirement plan and timelines.
> ➤ Summaries of any data migration activities.
> ➤ Identification of the storage location of documentation relating to the system.
> ➤ Obsoleting of SOPs.

It must be stressed that GAMP is a set of principles, a set of guidelines that aim to achieve compliant computerised systems that are fit for intended use. GAMP guidelines differ to 21 CFR QSR regulations as they are not legal or statutory requirements. However, they represent industry best practice and complement the validation efforts that are legal requirements and statutory requirements.

Regulatory Review

Software validation is a requirement of the quality system regulation, 21 Code of Federal Regulations (CFR) Part 820. Validation requirements apply to:

(1) software used as components in medical devices,
(2) software that is itself a medical device, and
(3) software used in production of the device or in implementation of the device manufacturer's quality system.

Note: EU GMP Annex 11 provides information on the inspection of 'computerised systems'.

In addition, computer systems used to create, modify, and maintain electronic records and to manage electronic signatures are also subject to the validation requirements. Such computer systems must be validated to ensure accuracy, reliability, consistent intended performance, and the ability to discern invalid or altered records. The regulated user should be able to demonstrate through the validation evidence that they have a high level of confidence in the integrity of both the processes executed within the controlling computer system and in those processes controlled by the computer system within the prescribed operating environment.

Specification Hierarchy

An equipment URS can define the requirements of a computerised or automated system along with the operational, functional, process and safety specifications mandated by the customer. If the system is bespoke and complex, an SRS may be written to more clearly detail the software requirements, automation and functionality. Similarly, an FDS (Functional Design Specification) can be written to address how mechanical or physical processing occurs.

URS to SRS

Scenario: a URS is written to specify the requirements for an automated blister packaging line in a medical device company.
The URS details the following:

URS R1.0 – The machine shall be capable of operating in various modes to allow the manufacture of product and other debugging activity.

In turn, an SRS can be written to interpret the URS, for example, an SRS system shall have the following modes:

SRS 1.1 Run empty mode - in this mode the equipment does not accept any new product.
SRS 1.2 Production mode - every station operates within the machine.
SRS 1.3 Bypass mode - where any operation can be disabled.

Another two examples of a URS requirement being transposed to an SRS requirement are shown below:

Example 1: URS Requirement

URS R1.0 In the event of an E-Stop activation, all sequencers shall be maintained and shall retain the sequence step that they were in at the time of the E-Stop activation.

SRS Requirement

SRS 1.0 The lot count and lot integrity must be maintained after an E-stop activation.

Example 2: URS Requirement

URS R1.0 The system shall use password protection
SRS Requirement

SRS 1. 1 Basic machine functionality (cycle start/stop, fault reset, and manual operations) requires no security.

SRS 1.2 As required, user IDs will be assigned to security groups for authentication. Authentication will be via active directory authentication against domain accounts.

SRS 1.3 An auto-logoff feature shall be incorporated in the design.

Examples of Security Requirements:

- ➢ Three levels of access required; operator, engineer and maintenance
- ➢ Engineer - access to all screens, to modify process settings maintenance - access to functions required to perform machine maintenance activities.
- ➢ Operator - restricted access, does not have access to change process settings.
- ➢ Different access levels will require different passwords.
- ➢ No security will be required for basic operations (start/stop).
- ➢ A user auto-logoff feature shall be incorporated in the design. The auto- logoff time shall be configurable.
- ➢ A soft copy of program settings must be provided with delivery of the equipment.

Examples of Security Requirements:

- ➢ The reject count and yield must be displayed on the HMI screen.
- ➢ Real-time readings for all critical parameters shall be visible on the HMI screen.

> ➢ All critical parameters shall be adjustable via the HMI screen.
> ➢ The status of each door should be visible on the HMI screen.

Examples of EHS Requirements:

> ➢ Activated E-Stops shall be clearly displayed on the HMI screen with a suitable alarm message generated.
> ➢ Activated E-Stops shall result in no further movement of the system until the E-stop is reset and all alarms are cleared.

System Categorisation

GAMP 5 makes provision for four categories of software in order to distinguish the level of customisation/configurability that exists across software serving different functions.

GAMP Software Category 1, Operating systems
GAMP Software Category 2, Non-configured software
GAMP Software Category 4, Configurable software packages
GAMP Software Category 5, Custom software

GAMP Software Category 1, Operating Systems

Category 1, operating systems, covers established, commercially available operating systems. These are not subject to validation themselves, the name and version of the operating system must, however, be documented and verified during Installation Qualification (IQ). Application software hosted on operating systems need to be validated.

GAMP Software Category 2, Non-Configured Software

Category 3 covers commercially available, standard software packages and "off-the-shelf" solutions for certain processes. The configuration of software packages should be limited to adaptation to the runtime environment (for example network and printer connections) and the configuration of the process parameters. The name and version of the standard software package should be documented and verified in an Installation Qualification (IQ). Special user requirements, such as security, alarms, messages, or algorithms must be documented and verified in an Operational Qualification (OQ).

GAMP Software Category 4, Configurable Software Packages

GAMP Software Category 4 (Configurable Software Packages Category 4) covers configurable software packages that allow special business and manufacturing processes. This involves configuring predefined software modules. These software packages should only be considered as belonging to Category 4 if they are well-known and mature. Normally, a supplier audit is necessary. If this is not available, the software packages should be handled as Category 5. The name, version, and configuration should be documented and verified in an Installation Qualification (IQ). The functions of the software packages should be verified in terms of the user requirements in an Operational Qualification (OQ). The validation plan should take into account the lifecycle model and an assessment of suppliers and software packages.

GAMP Software Category 5, Custom Software

GAMP Software Category 5, Custom Software Custom/Bespoke Software (GAMP Software Cat 5) is software that contains custom code designed or modified specifically for a particular customer. As the code is custom it presents a greater risk. This risk must be mitigated with the right approach to the validation.

GAMP Considerations

Correctly assigning a GAMP software category to equipment, a system or process is an important activity that should be completed early in the planning stage of a project. There must be some degree of familiarity with the equipment or system. The manufacturer or vendor can be a source of information that may help the designation. In many cases, companies create tools or processes that help determine what GAMP software category applies. These have different names such as questionnaires, screening tools, planning tools etc.

Risk Assessments

A risk assessment process should be applied to cGxP computerised systems in order to identify and mitigate potential risks to (1) patient safety, (2) product quality and (3) data integrity. Results identified through a risk assessment help to determine the validation strategy, the effort and time required, and allow better targeting of the validation activities to the highest risks.

The risk assessment should be revised during the Software Development Lifecycle (SDLC) if the functionality, requirements or intended use of the system changes. The Risk Assessment activity should also be evaluated during system build-up as well as when implementing changes. Risk assessment tools for cGxP computerised systems are typically completed during the planning stage, specification stage and post qualification if a change or update is required.

Planning Stage

Initial Impact/Risk Assessment – during the planning phase to identify the level of impact and GxP relevance of the system/equipment. (Tools used: High Level Risk assessment).

Specification Stage

Functional or Quality Risk Assessment – during the specification phase - identify potential risks and possible mitigations to be to be introduced to the process. (Tools used: Quality Risk Matrix, (p)FMEA).

Changes to the System

Impact Assessment of changes – as part of the change control process in the system operational phase. The following diagram defines the risk assessment steps within the System Life Cycle (Tools used: Impact assessment checklist, change control procedures).

Quality Risk Matrix
A QRM is a risk assessment that identifies and manages the risk to patient safety, product quality and data integrity that relate to the system processes. Risk scenarios or potential causes should be developed for each identified function or process step and then assessed for the impact on patient safety, product quality or data integrity. Risk mitigations and controls should then be introduced to address both medium and high-levels of risk. The QRM requires three "assessments" in order to produce an estimation or overall risk (low, medium, high).

- Assess likelihood
- Assess detectability
- Assess severity

<u>Traceability Matrix</u>

A Traceability Matrix should be prepared as required in accordance with company and internal policy. It is also recommended by GAMP guidelines, ASTM E2500 and ISPE risk-based approach to validation. The matrix links the user requirements and specifications to the testing and validation activities. A traceability matrix illustrates that all user requirements are traceable to the verification/validation activity or vendor documents as relevant (FDS if applicable, design specifications etc. Generally, individual organisations will have an approved template to work from. However, the URS structure can form the basis of the template, with additional columns added to document the test/verification method, Reference documents (such as FDS and vendor specifications and design documents).

Configuration Identification

Software and hardware packages should be identified by a unique product identifier and a version number. For the software end-user, the parts of an automated system that are subject to configuration management should be clearly identified. The system should therefore be broken down into configuration items. These should be identified at an early phase of development so that a complete list of configuration items is defined and maintained. The application-specific items should have a unique name or version ID. The depth of detail when specifying the elements is decided by the needs of the system, and the organisation developing that system.

Requirements for the User ID and Password

User ID: The user ID of a system should have a minimum length agreed with the customer and should be unique within the system.

Password: A password should always consist of a combination of numeric and alphanumeric characters. When setting up passwords, the number of characters and a period after which a password expires should be stipulated. The structure of the password is normally selected to suit the specific customer. The configuration is described in the section security settings of password policy.
Criteria for the structure of a password are as follows:

- Minimum length of the password
- Use of numeric and alphanumeric characters
- Case sensitivity

Audit Trail

The audit trail is a control mechanism of a system that allows all data entered or modified to be traced back to the original data. A reliable and secure audit trail is particularly important in conjunction with the creation, change or deletion of GMP-relevant electronic records. In this case, the audit trail must archive and document all the changes or actions made along with the date and time. Typical contents of an audit trail must be recorded and describe the procedures; "who changed what and when" (old value/new value).

Uninterrupted Power Supply

An uninterruptible power supply (UPS) is a system for buffering the main power supply. If the power supply fails, the battery of the UPS supplies the required power. When the power supply returns, the UPS battery stops supplying power and is recharged. Some UPS systems provide the option of main power supply monitoring in addition to the buffering function. They guarantee an output voltage at all times without interference voltages. UPS systems are necessary so that process and audit trail data can continue to be recorded during power failures. The design of the UPS must be agreed with the system user and must be specified in the URS, FS or DS. The following points must be considered:

➤ Energy requirements of the systems to be supplied
➤ Power of the UPS
➤ Required duration of UPS buffering

The energy requirements of the systems to be buffered decide the size of the UPS. A further selection criterion is the priority of the systems. Systems with high-priority include:

➤ Automation system (AS)
➤ Archive server
➤ Operator station (OS) server
➤ Operator station (OS) clients
➤ Network components

Field devices that generally have relatively high energy requirements may also be included in the buffering depending on the power of the UPS. This must be decided in consultation with the system user and related to the classification of the process. Whatever is decided, it is important that the systems for logging data are included in the buffering. The time at which the power failure occurred should also be recorded. The use of UPS systems involves the installation of software. This should be installed and configured on the PC-based computers of the process control system to be buffered. The setup should also account for:

> Configuration of the power failure alarms
> Stipulation of the time before the PC is shut down
> Stipulation of the time during which UPS buffering is provided

Automation systems (AS) must be programmed so that the process control system changes to a safe state after a selectable buffer time if a power failure occurs.

Types of UPS

Due to the different requirements of the various devices involved, three classes have established themselves as stipulated by the International Engineering Consortium (IEC) in product standard IEC 62040-3 and the European Union EN 50091-3:

Offline UPS

The simplest and least expensive UPS systems (according to IEC 62040-3.2.20, UPS class 3) are standby or offline UPS systems. They protect only against power outages and brief voltage fluctuations and peaks. Undervoltage and overvoltage are not compensated for. Offline UPS systems switch to battery supply automatically if there is an overvoltage or undervoltage.

Line-Interactive UPS

The way in which line-interactive UPS systems (according to IEC 62040-3.2.18, class 2) function is similar to standby UPS systems. They protect against power outage and brief voltage peaks and can compensate for voltage fluctuations continuously using filters.

Online UPS

Double conversion or online UPS systems (according to IEC 62040-3.2.16, class 1) count as genuine power generators that continuously generate their own line voltage. Connected consumers are therefore supplied permanently with line power without restrictions. At the same time, the battery is charged.

Software-Source Code Review

For GAMP Software Categories 4 and 5, source code review is advised unless the supplier has evidence of the same available for review. As part of Good Automated Manufacturing Practices, reviews should be completed as part of the development lifecycle. If a source code review is not completed, a justifiable rationale should be documented in an applicable document such as a validation master plan.

Calibration

A key part of any qualification is to confirm that the equipment is fit for the intended purpose. Each piece of equipment will have a defined operating range. For example, an oven may have an operating range of 20°C to 100°C ±5°C. However, the process window may only require a temperature range of 30°C to 60°C. In this instance, a calibrated range of 20°C to 70°C would suffice. However, if the process window or the temperatures at which product was manufactured ranged from 20°C to 100°C this would present a problem as it falls outside of the equipment qualification range when the calibration tolerance is taken into account.

Deviations

A deviation can be simply described as an unintended event which causes a test or verification to fail to meet expected acceptance criteria. Each company or organisation should have a procedure detailing the management of deviations. It is critical that all deviations are identified, investigated and evaluated for their impact on product quality, the risk/impact to the patient and the impact on the qualification or validation. The basic components to a deviation are listed below:

➢ Deviation description - provides the page and section of the deviation and an overall description e.g. document generation error, operator error, machine crash etc.

➢ Potential impact on product - does the deviation impact the product?

➢ Potential impact on validation/qualification - will the validation have to be repeated in part or in full?

➢ Investigation - DMAIC, RCA, Fishbone Diagram, 5W

➢ Root Cause - what is the concluding root cause?

➢ Planned Resolution - what actions are required to be implemented?

➢ Deviation Resolution (Actions completed) - were all the actions in the planned resolution implemented? What is the final result? Have the actions been effective?

Requalification

Over the lifetime of a piece of equipment, the need to requalify may arise. Therefore, any proposed change to equipment or a process must be assessed to see if the validated state will be impacted. It is therefore critical to understand clearly the nature of the change(s). Some scenarios where requalification of equipment may be required include:

> Major equipment repairs
> Moving equipment
> Changes to the upper and lower operating limits of the equipment
> Upgrading of software
> Hardware upgrades or changes
> Changes in performance and/or defect levels

After assessing any proposed changes based on the reasons listed above, a determination of the level of requalification is required. This may be limited to a partial requalification (addendum) or it may require a full requalification.

Process Validation

This section provides an introduction to process validation for medical devices. Process validation is a statutory and regulatory requirement for the manufacture of medical devices. Per FDA 21 Code of Federal Regulations Process Validation is a regulatory requirement of Good Manufacturing Practices (GMP) for both pharmaceuticals (21 CFR 211) and medical devices (21 CFR 820). In addition to the regulatory drivers, process validation is a requirement in order to obtain certification to international standards issued by many notified bodies. (E.g. ISO 13485 Medical Devices - Quality Management Systems, ASTM E2500-Standard Guide for Specification, Design, and Verification of Pharmaceutical and Biopharmaceutical Manufacturing Systems and Equipment etc.)

Traditional and New Approaches to Validation

Historically, process validation involved the testing and verification of all aspects of a process. While this may seem appropriate, it must be understood that in order to test/verify all aspects of a process, for it to hold weight, this activity must be documented and recorded. In this respect, an "all aspects" approach to process validation can be burdensome to resources. The traditional approach largely used the V-Model which set out a sequence of deliverables that should be completed. The use of risk assessments was limited as all requirements of a system were tested and qualified.

In recent years, a risk-based approach has been increasingly endorsed by regulatory authorities and hence adopted by medical device manufacturers. One such standard is the ASTM E2500. As the title suggests, it is primarily used within pharmaceutical and biopharmaceutical industries; its principles and core approach can be adopted by medical device manufacturers also. ASTM E2500 was designed to make the implementation process for GMP systems and validation more cost-effective. It aims to achieve this based on scientific and risk-based principles, focusing on the risk to the patient. However, at just a five-page document, ASTM E2500 lacks the detail required in order to meet regulatory expectations. While different terminology and philosophies exist, they do not change the regulatory expectations relating to validation. Both approaches exhibit common elements which include:

> Good engineering practices
> Planning
> Requirements definition (URS etc.)
> Design review
> Change management
> Documented testing and inspection

While many manufacturers may predominantly choose a particular approach, it is common to see elements of both approaches (traditional and risk-based). Each individual company will shape its internal validation procedures to best suit its business needs of the company.

What Is Process - Operational Qualification (OQ-P)?

The ability of a process to produce product in accordance with pre-determined specifications under worst case conditions. PQ is only required if no worst case conditions are evident.

What Is Process-Performance Qualification (PQ)?

The ability of a process to consistently produce product in accordance with pre-determined specifications under anticipated conditions (normal/routine conditions). Before considering process validation in further detail, it is important to look at the pre-requisites and other supporting activities required. These are examined in the sections below.

Test Methods and Process Validation

It is important to consider test methods early on in the validation lifecycle. Before you can begin to consider process validation, test methods should be understood and in place.

A test method is a process or an action used to verify that a product feature meets a predefined specification. Tests methods can be physical or analytical in nature. Test method validation should be completed in advance of process validations to allow the proper assessment of process and product outputs meaning it is often a pre-requisite to process validation.

Examples of test methods include simple visual inspection by microscope, measurement of a dimension with a callipers or measurement of a dimension using an automated optical inspection system. Some test methods will involve MSA (Measurement System Analysis) studies, for example, a measurement of a dimension by an operator using a microscope. In contrast, a test method to determine organic residuals would require an analytical test method validation.

The equipment must be qualified (installation qualification and operational qualification) before the method is validated. Remember that testing completed in contract laboratories or specialist services also requires validation! Test methods are critical to the success and integrity of your process validation as they assess the outputs. E.g. what are the dimensions, physical attributes or chemical properties of the product and how do they conform to specifications?

Fundamentals of Process Validation

The most important point when it comes to validation is that validation is neither exploratory nor investigative. Equally, it is not an engineering study. If you are ready to validate a system or process, all of the groundwork must be completed. This means critical parameters must be defined and documented, with technical rationale on why such parameters are critical etc. This body of work is typically done during a process development study or protocol. Process validation is confirming that a process is capable of consistently manufacturing product under anticipated conditions. Remember that validation should be representative of the commercial process, so any issues in process validation will be repeated in commercial manufacturing.

Consistency, a core principle of process validation, is typically demonstrated by producing three batches/runs for a Process Performance Qualification (PPQ). These batches should be representative of normal production, i.e. the size of the batch should be typical of commercial volumes. The PQ study should be executed at nominal conditions (often termed "anticipated conditions") essentially referring to a controlled environment. Controlled material and controlled parameters (CPPs) are required. Nominal settings should be selected for PQ.

Process Validation and Dominance Factors

The concept of dominance is a term used to describe the "influential" or "dominating" effect on a system or process. Typical examples include the injection moulding process and packaging process. For example, an injection moulding process can be said to have material as a dominant factor. Batch-to-batch differences of resin or raw material may cause a change to outputs such as dimensions of a product or component. If dominant factors cannot be identified or understood a "Designed Experiment" (DoE) technique can be used to properly determine them. Dominance can be categorised into 5 sections: (1) Setup Dominance (2) Time Dominance (3) Worker Dominance (4) Information Dominance and (5) Component Dominance.

Setup Dominance

Setup Dominance - The process or equipment relies principally on a procedure or process setup. Process should be stable one "set-up".

Examples include ovens and package sealers. With regard to the oven, the setup would generally be controlled by a recipe or program. This program would be selected by the operator through the Human Machine Interface (HMI). The setup with the correct version of the recipe that contains the desired temperatures, times and pressures is therefore a critical input to the process. With regard to the packaging machine (blister packaging), the correct setup for the tooling and program is critical. If setup dominance is significant, it is best practice to have three separate set-ups/changeovers in the Performance Qualification (PQ).

Time Dominance

The process or equipment is subject to changes over time (drift over time in temperature, solvent cleanliness, tool wear etc.). The process may need a schedule of process checks and adjustments to ensure process consistency.

Examples include CNC machinery (tool wear) or aqueous-based cleaning systems. The tool may only be able to manufacture 1000 parts before defects or quality issues are encountered. If time dominance is significant, three time points or cycles of expected variation should be made e.g. three points in the cycle (start, middle and end) or three points in a shift (start of shift, middle of shift and end of shift).

Worker Dominance

For worker dominance, the process requires operator experience and skill. Examples include manual or hand finishing. If dominance is significant, ensure there are a minimum of three operators involved in the manufacturing/activity.

Information Dominance

With information dominance, the process or equipment requires the transmission and/or analysis of information. Examples include LIMS, MRP and ERP systems. A minimum of three information transmissions in the PQ should be completed.

Component Dominance

The process is influenced by the variability of the input materials and/or components. It requires robust inspection and sorting procedures as well as process adjustments. When component dominance is significant, ensure there are a minimum of three component/raw material batches in the PQ sampling plan. If component dominance is significant this can be mitigated by including the material/component variation in "worst case" testing as part of the Operational Qualification Process (OQ-P)

Process Operational Qualification (OQ-P)

During the Operational Qualification-Process (OQ-P) study, worst-case process conditions are normally employed. This may be worst case temperatures, speeds, feeds etc. The OQ-P should challenge the manufacture/processing of product at the limits of the processing window (process range). If no worst-case conditions exist, then an OQ may not be required and only a performance qualification is necessary.

A family or matrix approach is often used where similar products are to be validated. A particular product size of product configuration may be selected to represent the worst-case product. Therefore, by qualifying the worst case, all other products within that family of products would be considered validated. However, this approach must be clearly documented and technical rationale provided in advance of any qualification activities. This can be addressed in a validation plan or within a protocol.

Protocol Approval Checklist

The validation protocol is the means by which objective evidence is documented and gathered. The validation protocol is therefore a critical document. It should clearly set out the approach to the validation, detailing methods, tests and verifications to be completed and the acceptance criteria that applies to such tests and verifications. Remember, a validation document is a legal and regulatory document and can be subject to detailed scrutiny. Below are some suggested general checks to apply when writing validation protocols.

Author
- SOP available - Protocol conforms to validation procedure.
- Ensure item numbers and batch size are correct.

-Test methods are correct.

Sunject Matter Expert Reviewer
- Is the protocol number correct?
- Review content of protocol for accuracy and completeness.
- Protocol conforms to validation procedures.
- Procedure and evaluation table are appropriate and correct.

Engineering:
- Review content of protocol for accuracy and completeness.
- Specifications and operating parameters are correct.

QC / Laboratory
- Review content of protocol.
- Raw material specifications are in place.
- Finished product specifications are in place.
- Testing and sample size are correct.

Quality
- Review content of protocol.
- Protocol conforms to SOPs.
- Evaluation and acceptance criteria are appropriate.

Process Performance Qualification

The purpose of the PPQ is to demonstrate the capability of the process to consistently manufacture product to pre-determined specifications under normal operating conditions and defined parameters. Validation is confirmation, so process validation is confirming that a process is capable of consistently manufacturing product under anticipated conditions.

> Lots should be produced consecutively (in sequence)
> Lots must meet the acceptance criteria set out in the protocol
> The lot size should be reflective of the intended lot size and also take into account normal variation
> If a family approach or matrix approach is used, the product selection must be clearly justified and documented
> Execute under anticipated conditions (essentially this refers to a controlled environment)
> Controlled material, controlled parameters (CPPs)
> Nominal settings should be selected for PPQ

Yield Data (aka Process Yield Data)

Process yield is a term used in manufacturing to represent the overall process performance. Yield is most often expressed as a percentage of good/passing product. It reports the % of compliant units, that is units or products that meet the product acceptance criteria (eg. CQAs). The remaining "bad" units are classified as defects or scrap. In some manufacturing processes, rework is possible or permitted.

Yield data often forms part of the acceptance criteria for a validation. The overall process yield for each batch should be calculated and compared to the starting process weights or units to determine loss due to processing as it is common to lose material during processing.

Continued Process Verification

Once the initial validation is completed it is important that the system or process remains within the validated state, meaning that the system remains in a state of controlling process systems that capture information and data about the performance of the process. The use of statistical trending techniques should be considered. Data analysis of process and product should also include trending of raw materials, components and finished product. The purpose of process monitoring is to ensure critical parameters remain within control limits. It also helps to identify increasing variability or instability within the process which can then be investigated. All processes must have an upper and lower limit. If a process parameter only has a one-sided limit, then provide rationale in the OQ protocol to justify why a one-sided parameter window is acceptable. This requirement is not applicable to parameters that are set points.

Revalidation (or Maintaining a Validated State)

Revalidation is sometimes required if the original validation is no longer valid or representative of the process. Some instances where revalidation must be considered include changes to the process that can affect the product quality or efficacy, a removal or the addition of a processing step or transfer of the equipment to a different location. In many companies an impact assessment is conducted if there is a proposal to modify a manufacturing process. Some changes may not require any validation while others may require a verification run.

When changes are proposed to the validated state of a process, the proposed changes must be fully understood in terms of the impact to product quality and the validated state. A risk assessment should be conducted to determine risks and appropriate mitigations.

Scenarios on Maintaining a Validated State

Line Addition Product (New Product or Product Transfer)

This may be required if a new product has been introduced but uses the same process(es) for manufacturing. Typically this can apply if a new size has been introduced. For example, a new size of surgical blade.

Line Extension

A line extension commonly refers to a scenario where the product/process is different or considered outside the existing range or processing parameters.
Note: The impact on the validated state for line additions and line extensions should be assessed formally and documented. A line extension may require a new validation or addendum to the existing validation, whereas a line addition to add a new product may be within the scope of existing validations.

Acceptance Criteria

The acceptance criteria contained in validation protocols are normally based on established product specifications. For example, a contact lens manufacturing company may produce a lens with a diameter of 18mm ±0.2mm. The product produced during a process validation must be inspected to record the diameters of lenses being manufactured. Disposition of product is based on the product specification and determines if the product feature measured receives a pass or fail.

In addition to product specifications, it is common to have acceptance criteria such as yield, and OEE. The acceptance criteria for these conditions are normally driven by an internal company procedure or alternatively can be detailed in the validation plan or protocol study.

Validation Strategies

A family approach (a.k.a. Bracketing, Matrix Approach) to validation is often used where a variety of similar products are manufactured using the same equipment. For a process validation, a product that is representative of the family or group of products may be selected. Alternatively, a 'worst case' product may be selected as it presents the greatest challenge to manufacture to product specifications.

Principles of Worst Case Selection

Worst case is a particular condition, set of conditions, and/or set of process parameters generally made up of processing limits. Worst case conditions present the greatest chance of process issues or the greatest chance of failures due to product quality. Worst case conditions are used at the OQ-P stage to provide the greatest level of challenge, however, this is outside of normal operating conditions.

Change Control

Change control can be defined as a formal system in which qualified representatives of appropriate disciplines review changes that may affect the validated status of facilities, systems, equipment or processes and determine the need for action that would ensure and document that the system is maintained in a validated state. The term change management is also widely used. It can be defined as the systematic approach to proposing, evaluating, approving, implementing and reviewing changes. (Ref: ICH Q10)

Note: WHO guidance and PIC/s does not specify any particular requirements in regard to change control.

Impact of Changes

Changes in processes, materials and equipment are unavoidable throughout the lifecycle of a product or process. However, in order to ensure changes do not have an adverse impact on product safety or quality, proposed changes should be assessed by all stakeholders, typically involving review of the change by cross-functional teams with a documented review of the change along with any tests or verifications required in order to implement the change safely.

Tools in Identifying Potential Impact

6M/6P/4S are simple and effective tools to use in order to identify equipment, areas, processes or documentation that may be impacted by a planned change.

6 M	6 P	4 S
Machines	People	Surroundings
Methods	Process	Suppliers
Mother Nature	Policy	Systems
Materials	Plant	Skills
Measurement	Programs	
Manpower	Products	

Figure: Lean tools

Templates can also be used to visually document 6M or 6P methodologies.

Complaints and Recalls

Patients and medical professionals are encouraged to be vigilant and report any potentially defective or suspect products. Manufacturers must have a system in place to receive complaints and a written procedure that details how complaints are reviewed and acted upon. Manufacturers must also have a system that facilitates the recall of products known or suspected to be on the market.

Suspect or defective product

Complaint received

Assessment of complaint according to written procedure

Decision making

Recall (if required)

General points:

> ➤ Record all details of complaint

> ➤ Document the investigation

> ➤ Document all decisions

➤ Review type and amount of complaints regularly

➤ Communicate with competent authorities as appropriate

<u>Investigation</u>

Q7 Guidance states the following with regard to investigations:

➤ *The information reported in relation to possible quality defects should be recorded, including all the original details. The validity and extent of all reported quality defects should be documented and assessed in accordance with quality risk management principles in order to support decisions regarding the degree of investigation and action taken.*

➤ *If a quality defect is discovered or suspected in a batch, consideration should be given to checking other batches and in some cases other products, in order to determine whether they are also affected. In particular, other batches which may contain portions of the defective batch or defective components should be investigated.*

➤ *Quality defect investigations should include a review of previous quality defect reports or any other relevant information for any indication of specific or recurring problems requiring attention and possibly further regulatory action.*

➤ *The decisions that are made during and following quality defect investigations should reflect the level of risk that is presented by the quality defect as well as the seriousness of any non-compliance with respect to the requirements of the marketing authorisation/product specification file or GMP. Such decisions should be timely to ensure that patient and animal safety is maintained, in a way that is commensurate with the level of risk that is presented by those issues.*

➢ *As comprehensive information on the nature and extent of the quality defect may not always be available at the early stages of an investigation, the decision-making processes should still ensure that appropriate risk-reducing actions are taken at an appropriate time-point during such investigations. All the decisions and measures taken as a result of a quality defect should be documented.*

➢ *Where human error is suspected or identified as the cause of a quality defect, this should be formally justified and care should be exercised so as to ensure that process, procedural or system-based errors or problems are not overlooked, if present.*

➢ *Appropriate CAPAs should be identified and taken in response to a quality defect. The effectiveness of such actions should be monitored and assessed.*

➢ *Quality defect records should be reviewed and trend analyses should be performed regularly for any indication of specific or recurring problems requiring attention.*

Recalls

Recalls should be managed and co-ordinated by a responsible person with adequate support from a wider team to handle all aspects of the recall or complaint. This responsible person typically is independent of the sales and marketing organisation.

CHAPTER 2

Preparation for Audits

Introduction

Every company should strive to always be 'audit ready'. Regulators can perform 'unannounced' audits at any time. First impressions count and it's important to convey to the auditors that you have your facility, quality management system (QMS) and manufacturing process under control.

A key to success during any audit is preparation. You may like to consider establishing a company procedure for the management of GMP audits – from the opening to closing meeting. This should also include the steps to follow for an 'unannounced' audit.

Your procedure should define roles and responsibilities of all personnel likely to be involved in an audit including:

- security and reception
- escorts
- scribes (also known as note takers)
- subject matter experts (SMEs)
- runners (document retrieval)
- Scrub team (ensure documents are free of post-its and are approved)

Documentation

Prior to the audit, ensure that your team has reviewed any documentation that an auditor is likely to request. Make sure that all documentation is accessible and that all circulated documents are 'controlled copies' and up-to-date. Some documents, such as the procedure for handling deviations, OOS, CAPA, change control and release of product are viewed by the TGA as key procedures. Make sure that these are current and detail adequately your processes and controls.

A new requirement for Annual Product Review (APR) will be audited from the 1st July 2010. This is considered by the TGA one of the most important documents and indeed, you should also consider it as such. Be mindful, that the information from APRs and the deviation and change control registers (logs) will be used by the TGA auditors to focus their audit activities. Typically, the auditor will review the following documents before arriving on your site:

> - Site Master File (if the auditor is new to your site)
> - Validation Master Plan
> - previous audit findings and your responses
> - Complaints and adverse events.

Site Master File and Validation Master Plan You should review the content of your Site Master File (SMF) and Validation Master Plan (VMP) when you have been notified of an upcoming audit. If appropriate, the updated SMF should be sent to the auditor prior to the audit.

A 6M Approach to Audit Preparation

It is always good practice to adopt a methodology or a process based approach. In the days and weeks prior to an audit or site inspection, there any many areas that should to be reviewed for compliance and adherence to procedures and processes.

6M

- Machine (equipment, technology)
    - Equipment is validated, calibrated and maintained
    - SOPs and work instructions are available
- Method (process)
    - Test methods are qualified and validated
    - Written procedures are in place, followed and have adequate information.

- Material (includes raw material, consumables, and information)
    - Are materials approved prior to use?
    - Is there a quarantine process?
- Man
    - Personnel are trained and qualified
    - Job descriptions are in place and responsibilities are clearly defined

- Measurement (inspection, environment)
    - Measurement systems and qualified and validated
    - Measurement systems are fit for the intended purpose
    - Measurement systems are maintained and calibrated

CHAPTER 3

Inspection of Quality Systems

Introduction

The Top-down approach

A "top-down" approach to audits and inspections involves looking at the "systems" for addressing quality compliance prior to the examination of specific quality problems or issues.

The process for performing subsystem inspections is based on a "top-down" approach to inspecting and is typical of competent authorizes and notified bodies worldwide.

The "top-down" approach via subsystem review evaluates whether a company has addressed the basic requirements in that subsystem by defining and documenting appropriate procedures. This is followed then by an analysis of whether the procedures have been implemented and are been followed.

The Bottom-up Approach

A bottom-up approach to inspection involves looking a individual quality issues or non-conformances and then work back up through the quality system. The advantage of this approaches it that is facilitates focusing in on specific problems and evaluating the response and actions relating to problems or quality and compliance issues.

Key Subsystems of a Quality Management System

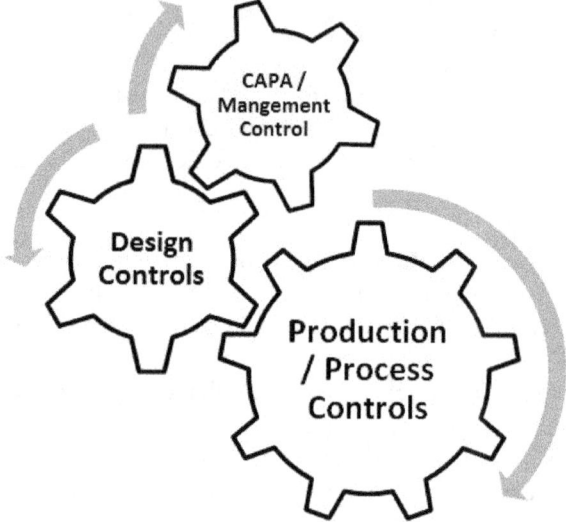

Figure: Key elements of a Quality Management System

Subsystems form the foundations of a quality system. Key subsystems include Management Controls, Corrective and Preventive Actions (CAPA) Design Controls and Production and Process Controls (P&PC). A subsystem inspection approach focuses on the important and critical aspects of quality system regulation. Both "top-down" and "bottom-up" inspectional approaches involve record review.

With the "topdown" approach, auditors sample records in many of the sub-systems to verify compliance to procedures and relevant regulations. The Quality System Regulation (21 CFR 820.3(k)) defines "Establish" as "define, document (in writing or electronically), and implement".

The Quality System Inspection Technique adopted by the FDA uses the "establish" approach in conducting inspections.
For each subsystem, an auditor will first determine if the company has defined and documented the requirements (CAPA, Design, etc.) by looking at procedures and policies.

Secondly, they typically examine records and documentation to determine if the company is following procedures consistently and if procedures in themselves are adequate and robust.

Quality System Management Controls

The purpose of a management control subsystem is to provide adequate resources from new product introduction throughout the lifecycle of a product and process. This includes the:

➢ Design
➢ Manufacturing
➢ Distribution
➢ Maintaining the validated state
➢ Retirement

Management controls also ensure the quality system is functioning properly; monitor the quality system; and make necessary adjustments.

A quality system that has been implemented effectively and is monitored to identify and address problems is more likely to produce devices that function as intended or medicines that are safe and effective.

A primary purpose of the inspection is to determine whether management with executive responsibility ensures that an adequate and effective quality system has been established (defined, documented and implemented) at the company / organization.

Inspection Objectives

1. Verify that the following have been defined and documented

    ➢ a quality policy
    ➢ management review
    ➢ quality audit procedures
    ➢ quality plan
    ➢ quality system procedures and instructions

2. Verify that a quality policy and objectives have been implemented.

3. Review the organizational structure to confirm that it includes provisions for responsibilities, authorities and necessary resources.

4. Confirm that a management representative has been appointed.

5. Verify that management reviews, including a review of the suitability and effectiveness of the quality system, are being conducted.

6. Verify that quality audits, including re-audits of deficient areas, of the quality system are being conducted.

Quality Policy

Each company must have a written quality policy. The definition of quality policy is provided in the FDA Quality System Regulation or other published guidance documents.

It summarizes the overall intentions and directions of an organization with respect to quality. A clear quality policy with realistic objectives forms the starting point of any Quality management system. The principles and objectives within the Quality policy should be supported and reflected by the content and processes in methods, procedures and documentation.

It is important to note that staff and personnel are not required to memorize or recite the quality policy but need to be familiar with its key points and know where to find it.

Management Review

Management reviews are a key regulatory requirement and also ensure the quality system is effective and is given the right amount of attention. If the top management does not embrace the importance of the quality system, it is destined for trouble.

Each company or manufacturer needs to have written procedures in place for conducting management reviews and quality audits. Procedures should make reference to defined intervals or a schedule of audits.

Quality System Procedures and Instructions

Manufacturers must prepare and implement all activities, including, but not necessarily limited to the applicable requirements of the Quality System Regulation (or Quality Management System), that are necessary to assure the finished product meets all pre-determined specifications.

The "quality system" as specified in the FDA 21 CFR, Quality System Regulation refers to all activities previously referred to as "quality assurance".

Slightly different terms may be used by various manufacturers such as "quality control" or "GMP Control" or "quality assurance" instead of quality system. However, this is acceptable once all the provisions of the requirements are satisfied.

Verification Methods

Verification methods that help determine if a quality policy and objectives have been implemented.

➢ Auditors/inspections can ask to see where the Quality Policy is available. For example: Is it in the Quality Manual or another written procedures

➢ Is the Quality Policy displayed in the building? Typically the quality policy is displayed in reception areas where employees can view it readily.

➢ Determine whether personnel and employees are familiar with the quality policy by directly asking them.

➢ Verification of training records to show individuals have been trained in the quality policy and objectives.

➢ Determine if personnel involved in managing, performing or assessing work affecting quality are granted the required independence and authority to perform those tasks.

➢ Review the organizational structure to confirm that it includes responsibilities, authorities and adequate resources.

➢ Verify management reviews are been undertaken and review of the suitability and effectiveness of the quality system, are being conducted. Dates and finding of management reviews should be documented.

➢ Management reviews should be relatively frequent enough, not allowing a period of time which would result in them being out of touch with the issues and points of discussion within the company.

➢ Management review procedures or instructions should include a requirement that the results of the reviews be documented and dated.

➢ Management reviews should verify that quality audits, including re-audits of deficient matters, of the quality system are being conducted by the appropriate personnel.

➢ The audit procedure should be reviewed at a defined interval for effectiveness and to ensure it is current and covers all aspects of the facility or company.

CHAPTER 4

During the Inspection

Audit Objectives

It is important bear in mind the fundamental purpose of an audit or inspection. The actions of the auditor can be attributed to key principles that must be evident within a company. They include the following:

➢ Is the manufacturing site in a state of control
➢ Is the manufacturer compliant with regulations and local laws
➢ Is the manufacturer following its own procedures
➢ Are system failures evident
➢ What is the level of quality? Is defective and non-conforming product a re-occurring issue?

At the opening meeting the lead auditor typically covers the following points:

➢ introduction of the audit team
➢ confirmation of the audit scope and objectives
➢ presentation and brief discussion of the audit plan
➢ discussion of the methods and procedures to be used during the audit
➢ discussion of the communication links during the audit

> ➢ confirmation of the proposed schedule over the course of the audit
> ➢ establishing a time and date for the closing meeting

In similar fashion, during the opening meeting, the host site should cover some basic information such as:

> ➢ company policy on health and safety
> ➢ company policy on hygiene
> ➢ company policy on mobile/cell phones
> ➢ normal operating hours (e.g. 8am to 5pm)
> ➢ times for lunch, breaks
> ➢ Smoking/eating habits

Upon arrival at a plant, inspectors are required to present identification and credentials.

Usually, the investigator will examine your production process, look at certain records and collect samples.

At the conclusion of the inspection, the investigator will discuss with management any significant findings and concerns; and leave with your management a written report of any conditions or practices, which, in the investigator's judgment, indicate objectionable conditions, or practices.

This is referred to as "Inspectional Observations," or also known as an FDA Form 483. This is a useful document management can use as a basis for corrective action. The FDA will not usually recommend specific corrective measures.

Conducting Internal Audits

As part of Quality Management systems a firm or manufacturer must conduct regular audits and report the findings. Some guidance on conducting internal audits includes:

- ➤ The management team or a representative should be available during the audit
- ➤ Access to documents and records relevant to the audit scope must be granted to auditors during the audit
- ➤ The audit should include a site tour for first time auditors
- ➤ Audits should be performed following a procedure or company standard
- ➤ Findings must be noted during the audit and discussed as appropriate
- ➤ A follow-up of previous audit findings should be completed

Classification of Findings

This section provides a general guide to the classification and categorization of findings.

Minor

- ➤ Isolated issues that if left unresolved, would not cause risk to the patient or product, but indicate minor departures from GxP

➤ A finding which cannot be classified as critical or major but indicates a potential departure from GxP

If substantial numbers of minor findings or re-occurring findings are observed it could indicate the beginning of a trend in compliance issues or call in to question the state of control of a facility.

Major

➤ A quality system with multiple findings that are similar in nature
➤ Previous minor audit findings that have not been adequately addressed.

Critical

➤ An occurrence which is likely to present a risk to patient health
➤ An occurrence that represents a serious violation of legislation guidelines quality standards
➤ An occurrence of fraud such as falsification of a product or a piece of information

➤ Where previous major audit findings have not been adequately addressed
➤ A series of major findings highlighting a wider failure in the quality management system

CHAPTER 5

Biotechnology Inspection Guide

Inspection Basics

- ➢ Biotech inspections are product-specific
- ➢ A team (minimum of 2) typically conducts inspections
- ➢ Prior to inspection, the inspection team decides duties of each member
- ➢ Audits are not all inclusive and cannot cover validation data of all systems

Cell Culture and Fermentation

Master Cell Bank and Working Cell Bank

The starting material for manufacturing BDP includes the bacterial, insect or mammalian cell culture which expresses the protein product or monoclonal antibody that has a specific role in a treatment that is designed to benefit a patient.

To ensure genetic stability of the cell bank during storage and propagation is a major concern, it is important to know the origin and history (number of passages) of both the MCB and WCB.

A MCB ampule is kept frozen or lyophilized and only used once. Occasionally, a new MCB may be generated from a WCB.

Characterization and Qualifying Tests

As it is the starting point of the process, the MCB must be rigorously tested.

Typically, it is not necessary to test the WCB as extensively as the MCB and limited characterization of a WCB should suffice. The following tests are generally performed on the WCB.

➢ Phenotypic characterization

➢ Restriction enzyme mapping

➢ Sterility and mycoplasma testing

➢ Testing the reproducible production of desired product

Auditor /Inspection Focus

Verify that written procedures are in place document accurately what is submitted in the application/approval documents.

Determine that batch records follow written procedures.

Determine the identity/traceability of the MCB/WCB.

Examine the conditions of storage at each location.

Media

Media is designed to support the growth of microorganisms or cells. Different types of media are used for growing different types of cells. The raw materials used to prepare the media must be carefully selected to provide the proper rate of growth and the essential nutrients for the organisms producing the desired product.

Raw materials should not contain toxic components that may be carried through the cell culture, fermentation and the purification process to the finished product.

Water is an important component of the media and the quality of the water will depend on the recombinant system used, the phase of manufacture and intended use of the product.

Bovine Serum

Most mammalian cell cultures require serum for growth. Serum must be from a secure source and adequately tested to ensure it is not contaminated. Biological product manufacturers have been requested to determine the origin of these materials used in manufacturing. The media used must be sterilized. A sterilized in place (SIP) or a continuous sterilizing system (CSS) process can be used. Any nutrients or chemicals added beyond this point must be sterile. Air lines must include sterile filters to remove any microorganisms.

Auditor /Inspection Focus

Determine the source of serum

Confirm sterilization process and cycle has been properly validated to ensure that the media will be sterile

Verify raw materials have been tested by quality control and C of A issued

Determine the origin of all bovine material

Verify that expired raw materials have not been used in manufacture

Check that media and other additives have been properly stored according to specifications

Review of routine media fills

Culture Growth

Inoculation and Aseptic Transfer

Bioreactor inoculation, transfer, and harvesting operations must be done using validated aseptic techniques. This is to ensure patient safety and reduce the risk of contamination and batch rejection.

Containment

Bioreactor systems designed for recombinant microorganisms require not only that a pure culture is maintained, but also that the culture be contained within the systems.

For large-scale research or production, four physical containment levels are established:

> ➢ GLSP

> ➢ BL1-LS

> ➢ BL2-LS

> ➢ BL3-LS

GLSP

(Good Large-ScalePractice) level of physical containment is recommended for large-scale research of production involving viable, nonpathogenic and nontoxigenic recombinant strains derived from host organisms that have an extended history or safe large scale use. The GLSP level of physical containment is recommended for organisms such as those that have built-in environmental limitations that permit

optimum growth in the large scale setting but limited survival without adverse consequences in the environment.

BL1-LS

> *(Biosafety Level 1 -Large Scale) level of physical containment is recommended for large-scale research or production of viable organisms containing recombinant DNA molecules that require BL1 containment at the laboratory scale.*

BL2-LS

> *Level of physical containment is required for large-scale research or production of viable organisms containing recombinant DNA molecules that require BL2 containment at the laboratory scale.*

BL3-LS

> *Level of physical containment is required for large-scale research or production of viable organisms containing recombinant DNA molecules that require BL3 containment at the laboratory scale.*

> *No provisions are made at this time for large-scale research or production of viable organisms containing recombinant DNA molecules that require BL4 containment at the laboratory scale. Ref: FDA*

Extraction, Isolation and Purification

Extraction and Isolation

Ultrafiltration is commonly used to remove the desired product from the cell debris.

Isolation of materials is achieved via the principal of porosity. If the porosity of the membrane filter is calibrated to a specific molecular weight. It will allow molecules below that weight to pass through while retaining molecules above that weight that are not desired or harmful.

Centrifugation is also another method of isolation which is commonly used.

Purification: The purification process is primarily achieved by one or more column chromatography techniques.

➤ Affinity Chromatography

➤ Ion-Exchange Chromatography (IEC)

➤ Gel filtration

➤ Hydrophobic Interaction Chromatography (HIC)

➤ Reverse- Phase HPLC

Process Validation

FDA defined process validation in *"Guideline on General Principles of Process Validation"* as follows:

"Validation -establishing documented evidence which provides a high degree of assurance that a specific process will consistently produce a product meeting its pre-determined specifications and quality attributes."

Manufacturers should have validation reports for the various key process steps. For example, if depyrogenation tunnel is used to remove pyrogens (endotoxins). Evidence and data documenting that this process is consistently and effective should be available.

By determining endotoxin levels before and after processing, a manufacturer should be able to demonstrate the validity of this process. Typically endotoxins of a known quantity are used and a reduction post processing should be evident.

Validation

Validation of processes should be representative of the commercial process or normal production conditions. Sometimes a manufacturer will develop a process on a small scale at it is more feasible and cost effective. Scale-up studies can then be completed to increase batch sizes to commercial volumes when the small scale processes achieves its goals.

The process that is used to supply commercial volumes must be validated and the validation must address the intended size of the production size batch going forward.

Process Water/WFI

The intended use of the finished product must be considered when selecting water. Injectable products must utilize Water for injection. For in-vitro studies, purified water may be acceptable. Most companies manufacture WFI by reverse osmosis rather than by distillation. WFI systems for BDPs are the same as WFI systems for other regulated products. Where products are heat sensitive cold WFI is used for formulation. Cold systems are prone to contamination. Therefore, cold WFI systems should be monitored both for endotoxins and microorganisms. Auditors may request evidence of the ongoing monitoring of such systems.

Plant Environment

Microbiological quality of the environment during various processing steps is a concern. As the process continues downstream, increased consideration should be given to environmental controls and monitoring. The environment and areas used for the isolation of the BDP should also be controlled to minimize microbiological and other foreign contaminants. The typical isolation of BDP should be of the same control as the environment used for the formulation of the solution prior to sterilization and filling.

Cleaning

Validation of the cleaning procedures for the processing of equipment and vessels should be conducted. This is especially critical for multi-product facilities where there is frequent changes between products been manufactured. Limits on the amount of residual contaminates is important in ensuring the safety and efficacy of products. The manufacturer should determine the degree of effectiveness of the cleaning procedure for each BDP or intermediate used in that particular piece of equipment.

Validation data should confirm with documented evidence that the cleaning process will reduce the specific residues to an acceptable level. However, it may not be possible to remove absolutely every trace of material, even with a reasonable number of cleaning cycles. The permissible residue level, generally expressed in parts per million (ppm), should be justified by the manufacturer. Cleaning should remove endotoxins, bacteria, toxic elements and contaminating proteins.

Detailed Cleaning Procedure

A written equipment cleaning procedure that provides details of what should be done and the materials or chemicals are to be utilized. It is good practice to list the specific solvent for each BDP and intermediate. All solvents and chemical agents must go through an approval process to ensure suitability and incoming quality.

Sampling Plan

Post cleaning, routine testing should be completed to ensure any out of specifications or contamination issues are identified. The sampling approach should provide a degree of confidence that any contamination will be detected. Critical points or areas in the process can be targeted. Above all, the cleaning process and the related sampling approach must be proven via a validation study.

Analytical Method/Cleaning Limits

Residue limits established for each piece of apparatus equipment should be practical, achievable, and verifiable. Subject matter experts should be able to stand over limits based on technical and sound scientific rationale. This can be achieved by using data to monitor residual levels.

Processing and Filling

Processing

Typically BDPs cannot be terminally sterilized and must be manufactured by aseptic processing. The presence of process related contaminants in a product or device is chiefly a patient safety issue. Sources of contaminants include:

➢ cell substrate (DNA, host cell proteins, and other cellular constituents, viruses)

➢ the media (proteins, sera, and additives)

➢ purification process (process related chemicals, and product related impurities)

Auditors are particularly interested in examining media fill data and validation of the aseptic manufacturing process during inspections.

In-Process Quality Control

In-process testing is an essential part of quality control and ensures that the actual, real-time performance of an operation is acceptable.

Examples of in-process controls are: stream parameters, chromatography profiles, protein species and protein concentrations, bioactivity, bioburden, and endotoxin levels. This set of in-process controls and the selection of acceptance criteria require coordination with the results from the validation program.

Filling

The filling of BDP into ampules, cartridges or vials presents similar problems as with the processing of conventional pharmaceutical products.

For a new BDP facility, the journey of validating sterile operations, equipment and systems, can be a lengthy and requires skilled and experienced personnel.

Typically the product development and clinical effectiveness studies are done across different sites. This adds another layer of complexity for BDP companies.

The batch size of a BDP, at least when initially produced, likely will be small. Because of the small batch size, filling lines may not be as automated as for other products typically filled in larger quantities. Thus, there is more involvement of people filling these products, particularly at some of the smaller, newer companies.

Focus areas that often present risks include:

➤ Robustness of Environmental monitoring

➤ Technique regarding Aseptic manipulation

➤ Filling speeds

➤ Fill volumes

➤ Effect of oxygen exposure on product

➤ Contamination from manual handling

Laboratory Controls

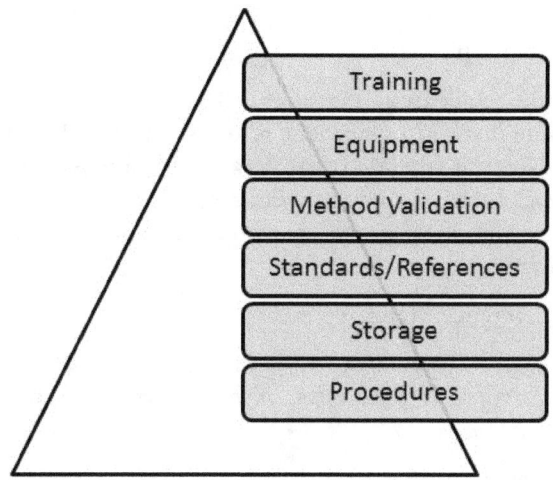

The below areas are subject to inspection and examination for compliance during audits:

<u>Training</u>

Laboratory personnel should be adequately trained for the jobs they are performing. This is achieved by training to SOPs/Work instructions on specific tasks and on-the-job training. Formal qualifications also provide a degree of competency and knowledge.

<u>Equipment Maintenance/Calibration/Monitoring</u>

Procedures and written records should ensure that a preventative maintenance schedule is in place for all equipment that is GxP. Calibration of critical gauges and instruments should be documented with each item re-calibrated at a defined interval.

Method Validation

Method validation ensures that not only test instruments and equipment is qualified, but that it is suited to the intended use. It also documents critical criteria such as equipment range, resolution and accuracy amongst other key parameters and acceptance requirements.

Standard/Reference Material

As per FDA guidance *"Reference standards should be well characterized and documented, properly stored, secured, and utilized during testing."*

Storage of Labile Components

Laboratory media, cultures and reagents, such as enzymes, antibodies, test reagents, etc., may degrade if not held under proper storage conditions. Supplier/manufacturer recommendations detailed in MSDS' of technical data sheets should be consulted to assist in meeting storage requirements.

Laboratory SOPs

Procedures should be written, applicable and followed.

Testing

The below tests are often relevant to intermediate or finished product, depending on the process and the intended use of the product.

Quality

1. Color/Appearance/Clarity

2. Particulate Analysis

3. pH Determination

4. Moisture Content

5. Host Cell DNA

Identity

Tests used to confirm identity must be validated. Availability of reference material should be checked. A comparison of the product versus reference preparation in a bioassay can be used to provide additional evidence relating to the identity and potency of the product.

Protein Concentration/Content

Some tests relevant to protein conc./content:

1. Protein Quantitations

2. Lowry

3. Biuret Method

4. UV Spectrophotometry

5. HPLC

6. Amino Acid Analysis

7. *Partial Sequence Analysis

Purity

"Purity" can be defined as the relative freedom from extraneous matter in the finished product.

This refers to the absence of residual moisture or pyrogenic substances. However, protein impurities are the most common contaminants. These may arise from the fermentation process, media or the host organism. Endogenous retroviruses may be present in hybridomas used for monoclonal antibody production.

Pyrogen Contamination

Pyrogenicity testing can be conducted by the limulus amebocyte lysate (LAL) assay.

In vivo (IN THE BODY) products be tested for pyrogens. Certain biological pharmaceuticals are pyrogenic in humans despite having passed the LAL test and the rabbit pyrogen test. It should be noted that some materials appear to be pyrogenic only in humans.

Pyrogen tests include:

- U.S.P. Rabbit Pyrogen Test
- Limulus Amebocyte Lysate (LAL)
- Pyrogen Assay

Viral Contamination

Tests for viral contamination need to be selected based on the cell substrate and culture conditions used. The purpose of the test is to demonstrate the absence of detectable adventitious viruses contaminating the final product.

Tests may include:

- Polymerase Chain Reaction (PCR)
- Cytopathic effect in several cell types
- Viral Antigen and Antibody Immunoassay
-

Microbial Contamination

Validated and appropriate tests should be conducted for microbial contamination. The purpose of testing is to demonstrate the absence of detectable bacteria (aerobes and anaerobes), fungi, and yeasts/molds.

Tests include:
- U.S.P. Sterility Test
- Heterotrophic Plate Count and Total Yeasts and Molds
- Total Plate Count
- Mycoplasma Test
- LAL/Pyrogen

Potency (Activity)

"Potency" can be defined as the specific ability or capacity of the product, as indicated by appropriate laboratory tests or by adequately controlled clinical data to produce a given result. Tests for potency should consist of either in vitro or in vivo tests, or both, which have been specifically designed for each product so as to indicate its potency.

Stability

"Stability" can be defined as the capacity of a product to remain within pre-defined specifications to ensure its identity, strength, quality, purity, safety, and effectiveness over the shelf-life of a product.

Testing might include stability of potency, pH, clarity, color, particulates, physiochemical stability, moisture and preservatives. Accelerated stability testing data may be used as supportive data. Accelerated testing or stress tests are studies designed to increase the ratio of chemical or physical degradation of a substance or product by using exaggerated storage conditions.

The purpose of stability testing is to determine kinetic parameters (heat, humidity) to predict the tentative expiration dating period. Stress testing of the product is frequently used to identify potential problems that may be encountered during storage and transportation and to provide an estimate of the expiration dating period.

Studies on the effects of temperature fluctuations encountered during shipping and transportation should be completed.

Batch To Batch Consistency

Manufacturers must be able to demonstrate lot-to-lot consistency of products manufactured so that they meet in process checks (as required) and release specification. Batch to batch consistency is demonstrated in process validation studies to ensure processes and the outputs are consistent and repeatable over time.

Auditor /Inspection Focus
(Laboratory Controls)

Personnel are trained to the tasks they complete

Equipment is calibrated, maintained and validated

All methods are validated and suitable for the intended use

Tests must be selected based on suitability for the particular product

Reference standards should be used where appropriate

Sampling size and frequency should be based on scientific rationale and supporting data

---

CHAPTER 6

Medical Device Inspection Guide

Introduction

The FDA issued the "Guide to Inspections of Medical Device Manufacturers" in the May 4, 1995 as part of its Compliance Program (CP). The Compliance Program contains information on when to do a directed inspection, the definitions of comprehensive and directed inspections, and other device specific policy requirements.

Overview

Other_useful sources of information relating to medical devices include:

➢ Compliance Program Guidance Manuals for Medical Device Manufacturers (CP 7382.830 (GMP)

➢ Investigations Operations Manual (IOM)

➢ Code of Federal Regulations, Title 21 (21 CFR) Part 820 Quality System Regulation

➢ Guideline on General Principles of Process Validation, FDA, May 1987

Similar to other sectors, auditors/investigators base their inspection on examining:

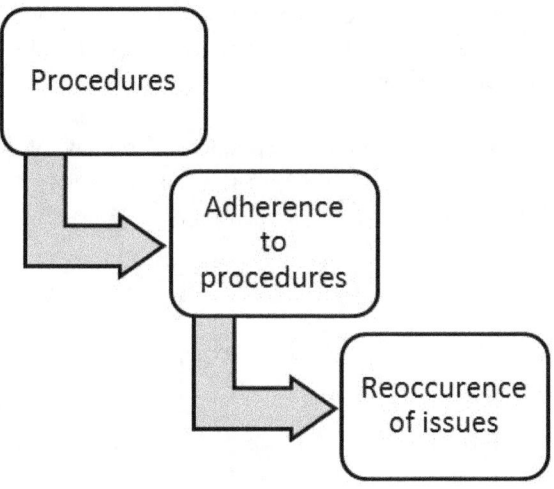

Figure: Auditor focus

These procedures, methods, records and reporting must be adequate. Rules of thumb auditors use to determine if procedures etc. are adequate include:

➢ the size of the company

➢ the complexity of the device(s)

➢ the relative risk to users if the device does not meet its product specifications

➢ the relative risk if the manufacturing facility is not operating in a state of control

Pre-Inspection Activity

Auditors complete due diligence prior to inspections by previous inspection findings, correspondence and any MDRs submitted or notifications of recalls.

In advance of an audit or inspection, a review of Medical Device Reports (MDRs) can be completed by both the inspector and host:

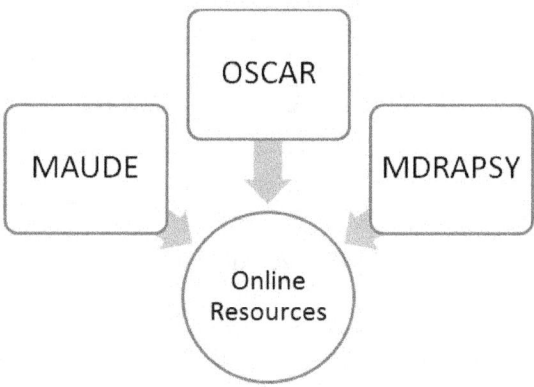

Figure: Online resources

The CDRH Information Retrieval System (CIRS) includes:

  ➢ Medical Device Reporting (MDR) data (MAUDE) Registration and Listing data and 510(k) and PMA summary data (OSCAR)
  ➢ MDRAPSY for MDR data prior to October 1996

MDR data that is most useful in preparing for an inspection of a medical device manufacturer includes specific MDRs for that manufacturer (i.e. query by firm's short name) for the time frame since the last inspection; or MDRs relative to the generic devices manufactured by that firm (i.e. query by product code) for some reasonable time frame. This data assists the investigator in determining possible problem areas in the manufacture, or design, of the device, or lot or batch specific issues. This data should be used to focus the inspection.

510(k) and PMA data assists the Investigator in determining what devices the firm is manufacturing and whether any new devices have been designed or changed since the last inspection. This data is useful in focusing the inspection on new or changed devices as well as devices that are higher risk devices, i.e. class II or III versus class I.

ISO 13485 Medical Device Directive

Introduction

ISO 13485 is the quality management standard of choice for manufactures of medical devices. Revised in 2016, ISO 13485:2016 "specifies requirements for a quality management system where an organisation needs to demonstrate its ability to provide medical devices and related services that consistently meet customer and applicable regulatory requirements."1 The scope of the standard can apply to any organisation or company involved throughout the life-cycle of a product, including design and/or development, production, storage and distribution, installation, or servicing of a medical device and design and development or provision of technical or professional services. 1 International Standards Organisation, www.iso.org

The recent revision is designed to address recent developments in quality management and other updated regulations that relate to the industry. Improvements in the new version of the standard include broadening its applicability to include all organisations involved in the life cycle of the product, from the concept stage to end of life along with greater alignment with regulatory requirements and post-market surveillance including complaint handling.

ISO 13485:2016 is also used by suppliers or external vendors that provide QMS related management system- services. Requirements of ISO 13485:2016 are applicable to organisations regardless of their size and regardless of their type except where explicitly stated. Wherever requirements are specified as applying to medical devices, the requirements apply equally to associated services as supplied by the organisation. If any requirement in Clauses 6, 7 or 8 of ISO 13485:2016 is not applicable due to the activities undertaken by the organisation or the nature of the medical device for which the quality management system is applied, the organisation does not need to include such a requirement in its quality management system. For any clause that is determined to be not applicable, the organisation records the justification as part of their certification and quality management system.

The Process Approach

ISO 13485 is based on a process approach to quality management. A process is any activity that receives inputs and converts them to outputs. For an organisation to function effectively, it has to identify and manage numerous linked processes. Furthermore, many processes impact other processes or downstream processes. The application of a system of processes within an organisation, together with the identification and interactions of these processes, and their management, can be referred to as the "process approach".

Directives Versus Standards

When it comes to regulated industries such as medical devices, it is first important to be familiar with some common terms and definitions and what they really mean. This chapter examines some key terms that are applied widely and relate to regulated industries.

They include:

• Directives

- Standards

- Notified Body

- Competent Authority

Directives

Directives are legal requirements which must be met by manufacturers or other bodies within the industry. Directives are based on legislation and are issued at governmental level. It is important to note that standards such as ISO 13485 help companies meet the requirements set up in directives. (See harmonised standards below)

Standards

Standards are not always mandatory. However, they help manufacturers be compliant with directives/legislation.

They also represent the current and best practice in the field of study/industry. Harmonised standards are European standards prepared under a mandate from the European Commission, referenced in the official journal, and drafted so that compliance with their requirements relates to one or more essential requirements of the directive. These standards have special status because, when a manufacturer can show that their products meet the requirements of the standard, there is a presumption that the product conforms to the essential requirements of the directive that is covered by the standard.

What is a Competent Authority?

When it comes to medical devices, a competent authority is the legally delegated authority mandated to monitor compliance to directives and legal requirements within the industry. The competent authority has the power to grant and revoke licenses.

Example of Competent Authorities:

• FDA (Food and Drug Administration) CFR Code of Federal Regulations – U.S.

• MHRA (Medicines and Healthcare Regulatory Agency - UK

• HPRA (Health Products Regulatory Agency) - Ireland

• JPAL (Japanese Regulations for Medical Devices) – Japan

What Is a Notified Body?

A notified body is a certification organisation which the national authority (the competent authority) of a member state designates to carry out one or more of the conformity assessment procedures described in the annexes of the medical devices directives. The Medicines and Healthcare Products Regulatory Agency is the UK competent authority under the three directives.

Organisations and Institutions

Many of the common acronyms that are referenced in literature relate to various standard setting organisations and industry representatives. Some of the more common bodies are listed below:

ISO: Internal Organisation for Standardisation

IMDR (F): International Medical Device Regulators Forum

ASTM: American Society for Testing and Materials

GHTF: Global Harmonisation Task Force

Basic Definitions (Source: Annex IX of Directive 93/42/EEC)

Intended Purpose: Intended purpose means the use for which the device is intended according to the data supplied by the manufacturer on the labelling, in the instructions and/or in promotional materials. (Chapter I section 1 of Annex IX of Directive 93/42/EEC)

Transient: Normally intended for continuous use for less than 60 minutes.

Short Term: Normally intended for continuous use for not more than 30 days.

Long Term: Normally intended for continuous use for more than 30 days.

Invasive Devices: A device which, in whole or in part, penetrates inside the body, either through a body orifice or through the surface of the body.

Body Orifice: Any natural opening in the body, as well as the external surface of the eyeball, or any permanent artificial opening, such as a stoma.

Surgically Invasive Device: An invasive device which penetrates inside the body through the surface of the body, with the aid of or in the context of a surgical operation.

Implantable Device: Any device which is intended:

- to be totally introduced into the human body or,

- to replace an epithelial surface or the surface of the eye, by surgical intervention which is intended to remain in place after the procedure. Any device intended to be partially introduced into the human body through surgical intervention and intended to remain in place after the procedure for at least 30 days is also considered an implantable device.

Medical Device: means any instrument, apparatus, appliance, material or other article, whether used alone or in combination, together with any accessories or software for its proper functioning, intended by the manufacturer to be used for human beings in the:

- diagnosis, prevention, monitoring, treatment or alleviation of disease or injury.

- investigation, replacement or modification of the anatomy or of a physiological process.

- control of conception which does not achieve its principal intended action by pharmacological, chemical, immunological or metabolic means.

A medical device may be assisted in its function by the following means:

Active Medical Device: any medical device relying for its functioning on a source of electrical energy or any source of power other than that directly generated by the human body or gravity.

Active Implantable Medical Device: any active medical device which is intended to be totally or partially introduced, surgically or medically, into the human body or by medical intervention into a natural orifice, and which is intended to remain after the procedure.

Custom-Made Device: means any active implantable medical device specifically made in accordance with a medical specialist's written prescription which gives, under his responsibility, specific design characteristics and is intended to be used only for an individual named patient.

Device Intended for Clinical Investigation: any active implantable medical device intended for use by a specialist doctor when conducting investigations in an adequate human clinical environment.

Intended Purpose: means the use for which the medical device is intended and for which it is suited according to the data supplied by the manufacturer in the instructions.

Putting into Service: means making available to the medical profession for implantation.

Where an active implantable medical device is intended to administer a substance defined as a medicinal product within the meaning of Council Directive 65/65/EEC of 26 January 1965 on the approximation of provisions laid down by law, regulation or administrative action relating to proprietary medicinal products (6), as last amended by Directive 87/21/EEC (7), that substance shall be subject to the system of marketing authorisation provided for in that directive.

Where an active implantable medical device incorporates, as an integral part, a substance which, if used separately, may be considered to be a medicinal product within the meaning of Article 1 of Directive 65/65/EEC, that device must be evaluated and authorised in accordance with the provisions of this directive.

ISO 13485 & Regulations

In chapter 2 the special status of harmonised standards was described which allows companies meet the essential requirements of Directives. In Europe, EN ISO 13485:2013 helps companies meet the requirements of: Directive 93/42/EEC on medical devices. This harmonised standard gives companies the "presumption of conformity" to complying with directives.

EN ISO 13485 was published in February 2013 and harmonised in August 2013 to cover the three directives:

- 90/385/ECC– The Active Implantable Medical Devices Directive (AIMDI)

- 93/42/ECC – The Medical Devices Directive (MDD)

- 98/79/EEC – In Vitro Diagnostic MDD (IVDMDD)

In the United States, medical device manufacturers need to meet the requirements of 21 CFR Part 820 of FDA regulations. While ISO 13485 is not an actual requirement, many companies will seek certification to the standard to support the exporting of products. In Australia, it is a regulatory requirement for manufacturers of medical devices to meet the requirements of ISO 13485. In Canada, certification to ISO13485 is part of the regulatory requirements. The content of ISO 13485 is interpretive (not prescriptive) which gives a degree of scope in how the requirements are applied and met within a company. ISO 13485 provides both a sound and widely recognised basis in meeting regulatory compliance for medical devices. Based off ISO 19001 however, ISO 13485 is a standalone standard for medical devices.

ISO 9001 has requirements and themes relating to continual improvement and customer satisfaction. These have been modified for ISO 13485.

Main differences between ISO 9001 & ISO 13485:

- Customer satisfaction is changed to customer feedback

- Extra requirements regarding procedures for ISO 13485

- Extra requirements for records ISO 13485 (e.g. retention)

- Continual improvement is restricted to continual improvement of the quality management system

ISO 13485 has extra requirements required for regulatory bodies such as post production review and management of advisory events.

ISO 13485 and ISO/TR 14969

ISO/TR 14969 is a technical report that is used for guidance on the application and implantation of ISO 13485. It is recommended for those responsible for the role out of ISO 13485 within their organisation. The content of ISO/TR 14969 is based on several established organisations such as the GHTF, ISO and input from regulatory bodies.

Standard Overview

ISO 13485 has 8 Clauses or Sections which make up the structure of the standard.

Section 0 Normative References, Definitions and Terms

Section 1 Requirements of the Quality Management System (QMS)

Section 2 Normative References

Section 3 Terms and Definitions

Section 4 Requirements of the Quality Management System (QMS)

Section 5 Management Responsibility

Section 6 Resource Management

Section 7 Product Realisation

Section 8 Measurement, Analysis and Improvement

CLAUSE 1: SCOPE

This section refers to the scope and application of the standard.

The organisation must be able to show its ability to provide medical devices to meet customer requirements and regulatory requirements

A key aim of the standard is to allow harmonisation to regulatory requirements

The scope of the QMS must relate to medical devices for a company to be able to use ISO 13485.

Some examples of what's in scope of the standard include (1) the manufacture of hip implants, (2) the design and manufacturing of in-vitro blood testing devices, (3) contact analytical testing (4) consultancy services. The terms "where appropriate" and "if appropriate" are used throughout the standard, therefore, it should be met by the organisation unless a justification is documented.

CLAUSE 2: NORMATIVE REFERENCES

This clause states that when working with ISO 13485, refer to ISO 9000:2000 for fundamentals and vocabulary.

CLAUSE 3: TERMS AND DEFINITIONS

This clause provides terms and definitions. It is very useful in the early days of establishing and implementing ISO 13485 to ensure that terms and definitions are clearly understood.

CLAUSE 4: QUALITY MANAGEMENT SYSTEM

Clause 4 details the general requirements that relate to the quality management system, the documentation requirements and record requirements.

Clause 4 includes:

4.1 General requirements clause

4.2 Documentation requirements clause

CLAUSE 4.1 GENERA REQUIREMENTS

The organisation must implement a Quality Management System, or QMS in order to provide the framework and structure to achieve ISO 13485 roll-out and implementation. However, the role of the QMS does not stop there. After initial roll-out, the requirements of the standard must be maintained and determined to be effective on an on-going basis. The following processes should be documented:

- List of all processes

- Process interactions

- Monitoring of processes

- Resources to facilitate rollout of processes

- Measure and monitor effectiveness

- System of identifying improvements

CLAUSE 4.2: DOCUMENTATION REQUIREMENTS

When it comes to the regulated industries such as the medical device industry, every process and procedure must be documented. Documentation ensures that everyone is working in the same manner with the same procedures. However, documentation is more than just writing down procedures and processes. It is also concerned with how documents are controlled, how they are updated and how they are stored.

Electronic Document Management Systems

Electronic document management systems aka EDMS are now the norm and gold standard for most medium to large organisations. Many companies that provide medical device manufacturers with an EDMS can customise the system to match the business processes particular to an organisation.

With configurable or customisable software, validation and proper verification is important to ensure the system operates as intended. There are also regulatory requirements that stipulate the expectations and requirements of such systems. For example, the application of electronic signatures and the presence of audit trials. FDA 21 CRF Part 11 details the requirements with regards to electronic records and electronic signatures. For medicinal products in Europe, GMP V4 Annex 11 specifies similar requirements.

Changes and Updates to Documents

Revision control is a key element of the Quality Management Systems in place in regulated industries. As the need for changes in the document arises, the controlled document can be amended/updated. With each update the version number revises also. Some companies will use alphabetic revision control and to a lesser extent numeric revision control (Version A, Version B or Version 01, Version 02).

Controlled documents should always have a version number or revision number electronically on each page of the document. This is similar to books which always list what edition they are. e.g. first edition or second edition.

Records

Records are generated through the application of processes and procedures. These records can be related in quality inspection and manufacturing. The integrity and quality of records relating to the manufacture of medical devices is important, as it plays a part in safe-guarding the patient or user. Records may also help in the investigation of any quality issues, complaints or adverse events that may arise.

Principles of Good Documentation Practices or GDP, should be applied to records. In particular, handwritten entries should always be accompanied by a signature and date. This is important as traceability must be maintained in the event of an issue or complaint.

CLAUSE 5: MANAGEMENT RESPONSIBILITY

Clause 5 includes:

5.1 Management Commitment

5.2 Customer Focus

5.3 Quality Policy

5.4 Planning

5.5 Responsibility, Authority and Communication

5.6   Management Review

5.1 Management Commitment

It is essential that top management have an authentic and tangible commitment to meeting regulations and the expectations of customers. Quality should be at the forefront of all of activities. Management should encourage discourse and communication on all matters relating to internal processes, quality and the QMS as a whole.

5.2 Customer Focus

Customer –patient/user/doctor/family member

Customer feedback is a requirement of ISO 13485 and as such the manufacturer must engage with the customer. In instances where a defective product is received, the manufacturer must have a complaints process to facilitate proper feedback, communication and investigation.

5.3 Quality Policy

Simple statement /1 pager or more

Often quality policies will be displayed in reception areas etc. Copies should be signed and revision controlled.

Quality policy must have a commitment to maintain the effectiveness of the QMS.

5.4 Planning

Top management must plan quality objectives and ensure they are implemented and effective.

Some examples of quality objectives include :

reduce rework by 10%

reduce scrap by 5%

have customer complaints reduced by 2% per year

5.5 Responsibility, Authority and Communication

Roles and responsibilities are defined.

Job descriptions are in place.

Organisational charts are in place and accurate.

5.6 Management Review

The purpose of management review is to ensure the effectiveness of the QMS.

Inputs to management review include:

(a) Audit results

(b) Customer feedback

(c) Process performance and conformity

(d) Corrective and preventative actions

(e) Deviations

(f) Regulatory changes and revisions

CLAUSE 6: RESOURCE MANAGEMENT

Clause 6 of ISO 13485 is concerned with human resources, infrastructure and work environment.

Clause 6 includes:

6.2 Human resources

6.3 Infrastructure

6.4 Work Environment

People are the key part of any QMS. Therefore, they should have the appropriate level of education, skill and experience. A culture of quality must be lived by everyone.

People must be suitably trained. Training must be documented and consistent throughout an organisation.

Training must be seen to be effective. Proper records of education and training must be kept.

Human intelligence, human creativity and human labour are all key inputs to any factory or company manufacturing medical devices. Therefore, an organisation must be properly resourced in order to function correctly, meet the regulatory requirements and customer expectations.

6.2 Human resources

"Change the people or change the people"

With any organisation, it is only as good as the people it has in its make-up. Therefore, the people, operators, engineers, managers etc. all contribute to the quality management system. Clause 6.2.2 also specifies requirements with regards to competence, awareness and training. The person should be matched to the job in terms of their qualifications, experience and training. Typically, job descriptions are used to drive and capture these requirements. Nowadays, most multinational companies will ask for evidence of qualifications, training and experience. These documents are then held on file in the event of an audit. This is recommended practice for medical device companies. While the standard does not specifically call out the need to hold records of degrees and qualifications on file, the company or organisation needs to demonstrate the suitability of the person to their respective roles, and filing the qualification provides the easiest method.

6.3 Infrastructure

Infrastructure has the ability to impact the quality of products and services. Therefore, it must be fit for purpose. It is especially important if the organisation is involved with the manufacture of medical devices. The following element need to be considered with regards to infrastructure:

- Location of equipment and the operating environment

- Equipment installation and validation

- Utilities required for the operation of equipment and systems

- Layout of the factory – flow or raw materials, in-process materials and finished products

- Environmental systems such as HVAC and fire suppression systems

6.4 Work Environment

The work environment is also closely related to infrastructure within a given organisation and they can both affect or impact upon the quality of products manufactured.

Risk to product quality and patients is minimised by understanding the work environment and how it can impact the product. When the interactions and risks are understood, work can then be done to eliminate risks or at least control or monitor them. Environmental conditions that can impact upon product quality include:

- Humidity
- Temperature
- Air quality
- Room pressure differentials (negative / positive)
- Air flow/velocity

CLAUSE 7 : PRODUCT REALISATION

Clause 7 includes:

7.1 Planning of product realisation

7.2 Customer-related processes

7.3 Design and development

7.4 Purchasing

7.5 Production and service provision

7.6 Control of measuring devices

7.1 Planning of Product Realisation

Product realisation can be defined as a collection of processes and body of work that delivers a product or service to the customer. Remember, when it comes to medical devices, customers can be patients or users such as doctors and nurses. It should be noted that organisations can opt to exclude specific requirements, in cases where product realisation is not applicable. However, any such exclusion should be based on sound rationale with the case clearly documented. An example of this may be where design and development is not conducted by the manufacturer e.g. contract manufacturers.

7.1 Planning of Product Realisation

Planning is an often underestimated but remains a key element of product realisation. If adequate time and resources are given to planning, it makes all other processes run smoother, and therefore should help to produce improved products and services.

7.2 Customer-Related Processes

There are 3 elements that feed into customer-related processes. They include the following:

Determining the requirements related to the product Clause 7.2.1

Review of requirements relating to the product-Clause 7.2.2

Customer communication-clause 7.2.3

Customer requirements are typically captured in a User Requirements Specification. A requirements specification (URS) documents all of the desired attributes of a product or service. They can be made up by a combination of CQAs, regulatory requirements and design requirements. A URS can then form the basis for review of the product or service requirements.

With regard to customer communication, it is important to remind ourselves that we are concerned with ISO 13485 which as we very well know by now is the standard for medical devices. Therefore, having the right information available to the customer, patient or end user is important. When additional information needs to be transmitted or updates to information need to be communicated, an advisory note can be issued. Another important aspect of customer communication is customer feedback. This communication can be made up of positive feedback from the customer or users, or when there is a query with regard to a product or service. Therefore, processes or systems must be in place to make communication between customer and company both effective and timely.

7.3 Design and Development

Design and Development Verification and Validation ensure that the product is designed, developed and subsequently manufactured meeting all the customer requirements, regulatory requirements and business requirements. These requirements are classed as inputs to the design and development, and verification and validation ensure the inputs have been adequately taken into account.

The design and development testing sometimes replicate the commercial applications of the medical device, hence providing a realistic challenge in order to have confidence in the medical device.

Design Control

Design control is a necessary practice that ensures good engineering principles are maintained throughout the design phase of a product. It also refers to the continual design and development of the product through its very lifecycle. The design and development files and history must be controlled and maintained, with any changes properly assessed, tested and documented.

7.4 Purchasing

Bearing in mind that a quality management system considers all aspects of an organisation's functioning, purchasing and procurement of materials necessitates putting robust controls in place. Simply put, a purchasing process must exist.

7.5 Production and Service Provision

This requirement of ISO 13485 is an extensive section with a great deal of importance associated with it. As we are dealing with the manufacture of medical devices (or other activity associated with medical devices) there are specific requirements for sterile products. If a product is sterile, its use or application is likely to be associated with greater risks to the patient. Therefore, extra safeguards must be in place for sterile medical devices. Key sections of Clause 7.5 include: (1) control of production and service provision – both general and specific requirements, (2) specific requirements for sterile medical devices, (3) validation of equipment and processes for production and service provision, (4) traceability and identification, (5) preservation of product controls with regard to monitoring and measuring medical devices.

7.6 Control of Measuring Devices

This clause requires an organisation to identify what monitoring and measuring is required and to ensure the product or service meets the customer requirements. A calibration procedure must also be maintained to ensure the equipment is accurate and reliable. Calibration must ensure that:

Equipment used to verify product quality is calibrated to a periodic schedule.

- The calibration is performed to an international standard.

- The calibration status of the equipment is recorded and visible.

- The equipment must be located within a suitable area in order to maintain accurate and reliable results.

If an organisation uses any computer software to monitor or measure outputs, the software must be verified before use via the appropriate validation and qualification activities.

CLAUSE 8 : MEASUREMENT ANALYSIS

Clause 8 includes:

8.1 General requirements

8.2 Monitoring and measurement

8.3 Control of nonconforming products

8.4 Analysis of data

8.5 Improvement

8.1 General Requirements

Measurement, analysis and improvement are the key themes of clause 8. As with all medical devices, inspection and testing both during manufacturing and post manufacturing is necessary to ensure products and services function as intended and without defects. With any type of measurement or inspection analysis, the method used to complete the testing is critical. The method must be fit for purpose, and the equipment must be suitable. This "method validation" typically is done during the design and development phase.

8.2 Monitoring and Measurement

Monitoring and measurement are dependent on the information or feedback provided from various sources. The most important feedback is the post-production feedback that is gathered from customers or the end user. Again, this occurs over the whole lifetime of the product or service in question. There are a number of methods that can be used to obtain feedback. Some examples include:

-Customer surveys

-Customer complaints

-Review of regulatory databases such as MAUDE.

-Repair and servicing information

8.3 Control of Nonconforming Product

Non-conforming product presents a risk to patients or users of medical devices. When a situation arises where non-conforming product is manufactured or detected through inspection processes, the product must be controlled and segregated to prevent unintended use or distribution.

Some examples resulting in non-conformance are:

- When a manufacturing process drifts outside its validation window or operating parameters.

- A certificate of analysis for a raw material is not provided by the supplier or the results are out of specification.

- In-process testing was not completed at the defined intervals.

- Training of personnel completing tests is not current or is inadequate.

8.4 Analysis of Data

In any engineering activity, data and the quality of the data is a key factor in making the right decisions. Provided the data collected is relevant and accurate, analysis of data can provide important insights into process performance, quality control and product functionality. Data should be collated in a consistent way and controlled by a procedure. When it comes to medical device manufacturing, the sources and types of data are multiple. Data can be generated from in-process testing and data can be generated from end of line testing aka finished product testing.

8.5 Improvement

ISO 13485 fosters a culture of continual improvement. As we have seen, each activity can be described as a process. For example, a manufacturing process, a procurement process, a complaints process. The set of processes that make up the quality management system need to be continually reviewed to ensure they are suitable and effective for the task at hand. Typical tools used to keep improvement in mind include:

- Review of the quality policy and quality objectives

- Frequency and category of corrective and preventative actions (CAPA's)

- Customer complaints

- Management review input

CE Marking

In Europe a QMS is required for CE marking of a medical device that is placed on the market in the EU.

ISO 13485:2003 is a harmonised standard that can be used by companies to show conformity of their QMS to requirements of directives. EN ISO 13485:2012 was harmonised in August 2012. This allows compliant companies receive an EC Declaration of Conformity.

Summary of the CE Requirements

Manufacturers of class I devices or their authorised representatives must:

- review the classification rules to confirm that their products fall within class I (Annex IX of the Directive)

- check that their products meet the essential requirements (Annex I of the Directive)

- notify the competent authority, in advance, of any proposals to carry out a clinical investigation to demonstrate safety and performance of a device as required by the regulations

- obtain notified body approval for sterility or metrology aspects of their devices and where applicable prepare relevant technical documentation

- Draw up the 'EC Declaration of Conformity' (below) before applying the CE marking to their devices

- Register with the competent authority

- Implement and maintain corrective action and vigilance procedures including a systematic procedure to review experience gained in the post-production phase

- Make available relevant documentation on request for inspection by the competent authority.

In Europe, all medical devices must bear the CE marking of conformity (see Annex XII) of the directive) when they are placed on the market and/or put into service. The CE marking must appear in a visible, legible and indelible form on the device or its sterile pack, where practicable and appropriate, and where applicable on any instructions for use and sales packaging. For 'sterile' and 'measuring' devices, the CE marking must be accompanied by the identification number of the notified body that has acted under the relevant conformity assessment procedure.

EC Declaration of Conformity

In order to affix the CE marking, the manufacturer or their authorised representative must follow the EC declaration of conformity procedure referred to in Annex VII of the directive. This procedure must be completed prior to placing the device on the market. The 'EC declaration of conformity' is the procedure whereby the manufacturer or their authorised representative prepares the required technical documentation, puts into place corrective action and vigilance procedures and declares that the products meet the requirements set out in the directive.

Technical Documentation

The technical documentation should be prepared following review of the essential requirements and must cover all of the following aspects:

Description: A general description of the product, including any variants (for example names, model numbers and sizes).

Raw Materials and Component Documentation: Specifications including, as applicable, details of raw materials, drawings of components and/or master patterns and any quality control procedures.

Intermediate Product and Sub-Assembly Documentation: Specifications including appropriate drawings and/or master patterns, circuits, and formulation specification; relevant manufacturing methods and any quality control procedures.

Packaging and Labelling Documentation: Packaging specifications and copies of all labels and any instructions for use.

Design Verification: The results of qualification tests and design calculations relevant to the intended use of the product, including connections to other devices in order for it to operate as intended.

Risk Analysis: The results of risk analysis to review whether any risks associated with the use of the product are compatible with a high level of protection of health and safety and are acceptable when weighed against the benefits to the patient or user. If biocompatibility is relevant – for example for skin contact and invasive devices – a compilation and review of existing data or test reports based on the relevant standards is required.

Compliance with the Essential Requirements and Harmonised Standards: A list of relevant harmonised standards (for example sterilisation, labelling and information, biocompatibility, electrical safety, risk analysis, product group standards) which have been applied in full or in part of the products. If relevant harmonised standards have not been applied in full, then additional data will be required, detailing the solutions adopted to meet the relevant essential requirements of the directive. The manufacturer may choose to prove conformity with the essential requirements of the directive through the use of their own standards and/or other relevant published standards (ISO, EN, BS). However, the use of such standards does not give similar, immediate presumption of conformity to the essential requirements of the directive. Therefore, using a harmonised standard provides greater protection to the manufacturer.

Device Classification

The manufacturer, in preparing for CE marking, should first determine if their product falls within the scope of the directive or national regulation, either as a medical device or as an accessory to a medical device, as defined in Article 1 of directive 93/42/EEC and Article 2 of the regulation. In order to be classified as a medical device, the product should have a medical purpose and its primary mode of action will typically be physical.

Level of Risk

General medical devices and related accessories must be classified into one of four classes, which are based on the perceived risk of the device to the patient or user. The classification of a device determines the conformity assessment options that are applicable to the device, with higher risk devices undergoing higher levels of assessment.

Class	Risk level
I	Low Risk
IIa	Medium Risk
IIb	Higher Risk
III	Highest Risk

Classification Rules

There are eighteen rules outlined in Annex IX of the directive and related regulation that lay down the basic principles of classification. In MEDDEV 2.4/1 Rev. 8, these rules are further explained and descriptive examples are provided. The eighteen rules are subdivided into four groups as follows:

Rules	Device Type
Rules 1 – 4	Non-invasive Devices
Rules 5 – 8	Invasive Devices

Rules 9 – 12   Active Devices

Rules 13-18   Special Rule e.g. devices containing tissue of animal origin, drug-device combinations

Annex IX and related guidance documents outline a number of key characteristics, listed below, that must be considered to correctly classify a device using the eighteen classification rules:

General Principles of Device Classification

Medical devices are defined as articles which are intended to be used for a medical purpose. It is the intended purpose that determines the class of device and not the particular technical characteristics of the device. The intended purpose of the device should be substantiated (if required) and be representative of the technical characteristics of the device.

It is the intended and not the accidental use of the device that determines its class.

It is the intended purpose assigned by the manufacturer to the device that determines the class of device and not the class assigned to other similar products.

Accessories are classified separately from their parent device.

The mode of action of a medical device should be clear and evidenced with appropriate data to confirm this mode of action.

If the device can be classified according to several rules then the highest possible class applies.

Multipurpose equipment which may be used in combination with medical devices are not themselves classed as medical devices unless the manufacturer places them on the market with the specific intended purpose as a medical device.

If the device is not intended to be used solely or principally in a specific part of the body, it must be considered and classified on the basis of the most critical specified use.

Summary Of Rules

(Source: Guidelines Relating To The Application Of

The Council Directive 93/42/EEC On Medical Devices, MEDDEC 2.4/Rev.9 June 2010)

Rule 1

Rule 1: All non-invasive devices are in Class I, unless one of the other 17 rules apply. This is a fallback rule applying to all devices that are not covered by a more specific rule.

This is a rule that applies in general to devices that come into contact only with intact skin or that do not touch the patient.

Some non-invasive devices are indirectly in contact with the body and can influence internal physiological processes by storing, channeling or treating blood, other body liquids or liquids which are returned or infused into the body or by generating energy that is delivered to the body. These must be excluded from the application of this rule and be handled by another rule because of the hazards inherent in such indirect influence on the body.

Rule 2

Rule 2: All non-invasive devices are in Class I, unless one of the other 17 rules apply.

These types of devices must be considered separately from the non-contact devices of Rule 1 because they may be indirectly invasive. They channel or store substances that will eventually be administered to the body. Typically these devices are used in transfusion, infusion, extracorporeal circulation and delivery of anaesthetic gases and oxygen.

In some cases devices covered under this rule are very simple gravity activated delivery devices.

Rule 2: All non-invasive devices intended for channelling or storing blood, body liquids or tissues, liquids or gases for the purpose of eventual infusion, administration or introduction into the body are in Class IIa:

- if they may be connected to an active medical device in Class IIa or a higher class,

-if they are intended for use for storing or channelling blood or other body liquids or for storing organs, parts of organs or body tissues.

- in all other cases they are in Class I.

Rule 3

Rule 3: Non-invasive devices that modify biological or chemical composition of blood, body liquids or other liquids intended for infusion into the body.

These types of devices must be considered separately from the non-contact devices of Rule 1 because they are indirectly invasive. They modify substances that will eventually be infused into the body. This rule covers mostly the more sophisticated elements of extracorporeal circulation sets, dialysis systems and autotransfusion systems as well as devices for extracorporeal treatment of body fluids which may or may not be immediately reintroduced into the body, including, where the patient is not in a closed loop with the device.

Rule 3: All non-invasive devices intended for modifying the biological or chemical composition of blood, other body liquids or other liquids intended for infusion into the body are in Class IIb,

unless the treatment consists of filtration, centrifugation or exchange of gas or heat, in which case they are in Class IIa.

These devices (Rule 3) are normally used in conjunction with an active medical device covered under Rule 9 or Rule 11.

Filtration and centrifugation should be understood in the context of this rule as exclusively mechanical methods.

Rule 4

Rule 4: Non-invasive devices which come into contact with injured skin.

This rule is intended to primarily cover wound dressings independently of the depth of the wound. The traditional types of products, such as those used as a mechanical barrier, are well understood and do not result in any great hazard. There have also been rapid technological developments in this area, with the emergence of new types of wound dressings for which non-traditional claims are made, e.g. management of the micro-environment of a wound to enhance its natural healing mechanism.

More ambitious claims relate to the mechanism of healing by secondary intent, such as influencing the underlying mechanisms of granulation or epithelial formation or preventing contraction of the wound. Some devices used on breached dermis may even have a life-sustaining or lifesaving purpose, e.g. when there is full thickness destruction of the skin over a large area and/or systemic effect.

Dressings containing medicinal products which act ancillary to the dressing fall within Class III under Rule 13.

Rule 4: All non-invasive devices which come into contact with injured skin:

- are in Class I if they are intended to be used as a mechanical barrier, for compression or for absorption of exudates,

- are in Class IIb if they are intended to be used principally with wounds which have breached the dermis and can only heal by secondary intent.

Products covered under this rule are extremely claim sensitive, e.g. a polymeric film dressing would be in Class IIa if the intended use is to manage the micro-environment of the wound or in Class I if its intended use is limited to retaining an invasive cannula at the wound site. Consequently it is impossible to say a priori that a particular type of dressing is in a given class without knowing its intended use as defined by the manufacturer. However, a claim that the device is interactive or active with respect to the wound healing process usually implies that the device is in Class IIb.

Most dressings that are intended for a use that is in Class IIa or IIb, also perform functions that are in Class I, e.g. that of a mechanical barrier. Such devices are nevertheless classed according to the intended use in the higher class.

For such devices incorporating a medicinal product or a human blood derivative see Rule 13 or animal tissues or derivatives rendered non-viable see Rule 17.

Rule 5

Rule 5: Devices invasive with respect to body orifices.

Invasiveness with respect to the body orifices (ear, mouth, nose, eye, anus, urethra and vagina) must be considered separately from invasiveness that penetrates through a cut in the body surfaces (surgical invasiveness). For short term use, a further distinction must be made between invasiveness with respect to the less vulnerable anterior parts of the ear, mouth and nose and the other anatomical sites that can be accessed through natural body orifices.

Surgically created stoma, which for example allows the evacuation of urine or faeces, should also be considered as a body orifice.

Devices covered by this rule tend to be diagnostic and therapeutic instruments used in particular specialities (ENT, ophthalmology, dentistry, proctology, urology and gynaecology).

Rule 5: All invasive devices with respect to body orifices, other than surgically invasive devices and which are not intended for connection to an active medical device or which are intended for connection to an active medical device in Class I:

- are in Class I if they are intended for transient use,

- are in Class IIa if they are intended for short term use

except if they are used in the oral cavity as far as the pharynx, in an ear canal up to the ear drum or in a nasal cavity , in which case they are in Class I,

- are in Class IIb if they are intended for long term use,

except if they are used in the oral cavity as far as the pharynx, in an ear canal up to the ear drum or in a nasal cavity and are not liable to be absorbed by the mucous membrane, in which case they are in Class IIa.

All invasive devices with respect to body orifices, other than surgically invasive devices, intended for connection to an active medical device in Class IIa or a higher class, are in Class IIa.

Rule 6

Rule 6: Surgically invasive devices intended for transient use (< 60 minutes)

This rule primarily covers three major groups of devices: devices that are used to create a conduit through the skin (needles, cannulae, etc.), surgical instruments (scalpels, saws, etc.) and various types of catheters, suckers, etc.

This rule primarily covers three major groups of devices: devices that are used to create a conduit through the skin (needles, cannulae, etc.), surgical instruments (scalpels, saws, etc.) and various types of catheters, suckers, etc.

Rule 6: All surgically invasive devices intended for transient use are in Class IIa unless they are:

-intended specifically to control, diagnose, monitor or correct a defect of the heart or of the central circulatory system through direct contact with these parts of the body, in which case they are in Class III

-reusable surgical instruments, in which case they are in Class I

-intended specifically for use in direct contact with the central nervous system, in which case they are in Class III,

- intended to supply energy in the form of ionising radiation in which case they are in Class IIb,

- intended to have a biological effect or to be wholly or mainly absorbed in which case they are in Class IIb,

- intended to administer medicines by means of a delivery system, if this is done in a manner that is potentially hazardous taking account of the mode of application, in which case they are Class IIb.

Rule 7

Rule 7: Surgically invasive devices intended for short-term use (>60 minutes, <30 days).

These are mostly devices used in the context of surgery or post-operative care (e.g. clamps, drains), infusion devices (cannulae, needles) and catheters of various types.

Rule 7: All surgically invasive devices intended for short term use are in Class IIa unless they are intended:

- either specifically to control, diagnose, monitor or correct a defect of the heart or of the central circulatory system through direct contact with these parts of the body, in which case they are in Class III,

- or specifically for use in direct contact with the central nervous system, in which case they are in Class III,

- or to supply energy in the form of ionising radiation in which case they are in Class IIb,

- intended to have a biological effect or to be wholly or mainly absorbed in which case they are in Class III, - or to undergo chemical change in the body, except if the devices are placed in the teeth, or to administer medicines, in which case they are Class IIb.

Rule 8

Rule 8: Implantable devices and long-term surgically invasive devices (> 30 days). These are mostly implants in the orthopaedic, dental, ophthalmic and cardiovascular fields as well as soft tissue implants such as implants used in plastic surgery.

Rule 8: All implantable devices and long-term surgically invasive devices are in Class IIb unless they are intended:

- to be placed in the teeth, in which case they are in Class IIa,

- to be used in direct contact with the heart, the central circulatory system or the central nervous system, in which case they are Class III,

- to have a biological effect or to be wholly or mainly absorbed, in which case they are in Class III,

- or to undergo chemical change in the body, except if the devices are placed in the teeth, or to administer medicines, in which case they are in Class III.

- Directive 2003/12/EC introduced a derogation from this rule, reclassifying breast implants in Class III

Directive 2005/50/EC introduced a derogation from this rule, reclassifying hip, knee and shoulder joint replacements in Class III

Rule 9

Rule 9: Active therapeutic devices intended to administer or exchange energy.

Devices classified by this rule are mostly electrical equipment used in surgery such as lasers and surgical generators. In addition there are devices for specialised treatment such as radiation treatment. Another category consists of stimulation devices, although not all of them can be considered as delivering dangerous levels of energy considering the tissue involved.

Rule 9: All active therapeutic devices intended to administer or exchange energy are in Class IIa

unless their characteristics are such that they may administer or exchange energy to and from the human body in a potentially hazardous way, taking account of the nature, the density and the site of application of the energy, in which case they are in Class IIb. All active devices intended to control or monitor the performance of active therapeutic devices in Class IIb or intended to influence directly the performance of such devices are in Class IIb.

Rule 10

Rule 10: Active devices for diagnosis. This primarily covers a whole range of widely used equipment in various fields, e.g. ultrasound diagnosis, capture of physiological signals and therapeutic and diagnostic radiology.

Rule 10: Active devices intended for diagnosis are in Class IIa:

- if they are intended to supply energy which will be absorbed by the human body, except for devices used to illuminate the patient's body, in the visible spectrum,

- if they are intended to image in vivo distribution of radiopharmaceuticals,

- if they are intended to allow direct diagnosis or monitoring of vital physiological processes,

unless they are specifically intended for monitoring of vital physiological parameters, where the nature of variations is such that it could result in immediate danger to the patient, for instance variations in cardiac performance, respiration, activity of CNS in which case they are in Class IIb.

Active devices intended to emit ionising radiation and intended for diagnostic and therapeutic interventional radiology including devices which control or monitor such devices, or which directly influence their performance, are in Class IIb.

Rule 11

Rule 11: Active devices intended to administer and/or remove medicines, body liquids or other substances to or from the body. This rule is intended to primarily cover drug delivery systems and anaesthesia equipment.

Rule 11: All active devices intended to administer and/or remove medicines, body liquids or other substances to or from the body are in Class IIa, unless this is done in a manner:

- that is potentially hazardous, taking account of the nature of the substances involved, of the part of the body concerned and of the mode of application, in which case they are in Class IIb.

Rule 12

Rule 12: All other active devices. This is a fall-back rule to cover all active devices not covered by the previous rules.

Rule 12: All other active devices are in Class I

Special Rules 12-18

Rule 13: Devices incorporating, as an integral part, a medicinal product or a human blood derivative (See MEDDEV. 2.1/3 for further guidance).

Rule 14: Devices used for contraception or prevention of sexually transmitted diseases.

Rule 15: Specific disinfecting, cleaning and rinsing devices.

Rule 16: Devices to record X-ray diagnostic images.

Rule 17: Devices utilising animal tissues or derivatives.

Rule 18: Blood bags.

CHAPTER 7

Sterile Drugs Inspection Guide

Introduction

The manufacture of sterile drug products presents unique challenges. In particular the risk of microbiological contamination can jeopordise large batches of sterile bulk substances that is costly to manufacturers. Sterile bulk drug substances can be tested for microbial contamination which eliminates contaminated batches prior to further processing and in advance of getting to the patient. The safety and sterility of batches can also be called into question due to a lack of sterility assurance. It should be noted that in the manufacture of the sterile bulk powders there is no further processing of the finished sterile bulk powder to remove contaminants or impurities such as particulates, endotoxins and degradants.

Starting points of inspection:

➤ Microbiological contaminated batches, list (if any)
➤ Deviations
➤ Retested batches and reasons for same
➤ Rejected batches and reasons for same
➤ Environmental monitoring summary reports
➤ CAPAs /QA investigations

The manufacture of a sterile bulk substances usually include the following steps:

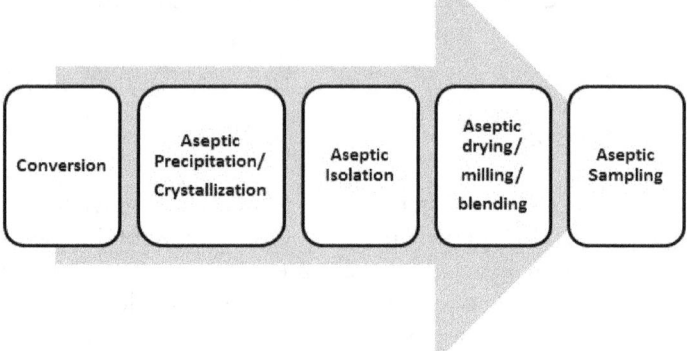

Figure: Steps in a Sterile Bulk Process

Conversion: non-sterile drug substances need to be converted to sterile form by dissolving in a solvent, sterilization of the solution by filtration and collection in a suitably sterilized vessel (crystallizer).

Aseptic precipitation/crystallization: processing of the sterile drug substance in the sterile reactor.

Aseptic isolation: isolation of the sterile substance by centrifugation or filtration/ultrafiltration.

Aseptic drying/ milling /blending: unit operations such as drying/milling/blending of the sterile drug substance.

Aseptic sampling: sampling and packaging the drug substance.

Components

In addition to the impurity concerns for the manufacture of bulk drug substances, there is a concern with endotoxins in the manufacture of the sterile bulk drug substances.

As with any operation, this may or may not be correct. For example, in an inspection of a manufacturer who conducted extensive studies of the conversion (crystallization) of the non-sterile substance to the sterile drug substance, they found no change from the initial endotoxin level. Organic solvents were used in this conversion. Thus, it is important to review and assess this aspect of the validation report.

Since endotoxins may not be uniformly distributed, it is also important to monitor the bioburden of the non-sterile substance(s) being sterilized. For example, gram negative contaminants in a non-sterile bulk drug substance prior to sterilization are of concern, particularly if the sterilization (filtration) and crystallization steps do not reduce the endotoxins to acceptable levels. Therefore, microbiological, as well as endotoxin data on the critical components and operational steps are typically reviewed during audits.

Facility and Plant

Facility and plant design for the aseptic processing of sterile bulk drug substances should have the same design features as an SVP aseptic processing facility. These would include temperature, humidity and pressure control. Because sterile bulk aseptic facilities are usually larger, problems with pressure differentials and sanitization have been encountered. For example, a manufacturer was found to have the gowning area under greater pressure than the adjacent aseptic areas. The need to remove solvent vapors may also impact on area pressurization.

Unnecessary equipment and/or equipment that cannot be adequately sanitized, such as wooden skids and forklift trucks, should be identified. Inquire about the movement of large quantities of sterile drug substance and the location of pass-through areas between the sterile core and non-sterile areas. Observe these areas, review environmental monitoring results and sanitization procedures.

Processing

Sterile powders are usually produced by dissolving the non-sterile substance or reactants in an organic solvent and then filtering the solution through a sterilizing filter. After filtration, the sterile bulk material is separated from the solvent by crystallization or precipitation.

Other methods include dissolution in an aqueous solution, filtration sterilization and separation by crystallization/filtration. Aqueous solutions can also be sterile filtered and spray dried or lyophilized.

With regard to processing of sterile product the following should be available for inspections:

> Engineering drawing of the piping system
> Validation reports of filtration and sterilization processes
> Information of filter specification and frequency of changes
> Procedures for integrity testing of filters

For spray drying of sterile powders, the source of air and the quality/specification, the chamber temperatures and the particle residence or contact time are important critical aspects. For bulk lyophilization, critical aspects include:

> air classification
> aseptic barriers for loading and unloading the unit
> uneven freezing
> partial meltback

Equipment

Equipment used in the processing of sterile bulk drug substances should be sterile and capable of being sterilized. This includes the crystallizer, centrifuge and dryer. The sanitization, rather than sterilization of this equipment, is unacceptable. Sterilization procedures and the validation of the sterilization of suspect pieces of equipment and transfer lines should be reviewed.

The method of choice for the sterilization of equipment and transfer lines is saturated clean steam under pressure. In the validation of the sterilization of equipment and of transfer systems, Biological Indicators (BIs), as well as temperature sensors (Thermocouple (TC) or Resistance Thermal Device (RTD)) should be strategically located in cold spots where condensate may accumulate.

SIP systems for the bulk drug substance industry require considerable maintenance, and their malfunction has directly led to considerable product contamination and recall. The corrosive nature of the sterilant, whether it is clean steam, formaldehyde, peroxide or ethylene oxide, has caused problems with gaskets and seals. In two cases, inadequate operating procedures have led to even weld failure. For example, tower or pond water was inadvertently allowed to remain in a jacket and was valved shut. Clean steam applied to the tank resulted in pressure as high as 1,000 lbs., causing pinhole formation and contamination. Review the equipment maintenance logs. Review non-schedule equipment maintenance and the possible impact on product quality. Identify those suspect batches manufactured and released prior to the repair of the equipment.

Another potential problem with SIP systems is condensate removal from the environment. Condensate and excessive moisture can result in increased humidity and increases in levels of microorganisms on surfaces of equipment. Therefore, it is particularly important to review environmental monitoring after sterilization of the system.

Environmental Monitoring

The environmental monitoring program for the sterile bulk drug substance manufacturer should be similar to the programs employed by the SVP industry. This includes the daily use of surface plates and the monitoring of personnel. As with the SVP industry, alert or action limits should be established and appropriate follow-up action taken when they are reached.

There are some bulk drug substance manufacturers that utilize UV lights in operating areas. Such lights are of limited value. They may mask a contaminant on a settling or aerobic plate. They may even contribute to the generation of a resistant (flora) organism. Thus, the use of Rodac or surface plates will provide more information on levels of contamination.

There are some manufacturers that set alert/action levels on averages of plates. For the sampling of critical surfaces, such as operators' gloves, the average of results on plates is unacceptable. The primary concern is any incidence of objectionable levels of contamination that may result in a non-sterile product.

As previously discussed, it is not unusual to see the highest level of contamination on the surfaces of equipment shortly after systems are steamed. If this occurs, the cause is usually the inadequate removal of condensate.

Since processing of the sterile bulk drug substance usually occurs around the clock, monitoring surfaces and personnel during the second and third shifts should be routine.

In the management of a sterile bulk operation, periodic (weekly/monthly/quarterly) summary reports of environmental monitoring are generated. Review these reports to obtain those situations in which alert/action limits were exceeded. Review the firm's investigation report and the disposition of batches processed when objectionable environmental conditions existed.

Validation

The validation of the sterilization of some of the equipment and delivery systems and the validation of the process from an endotoxin perspective have been discussed.

In addition to these parameters, demonstration of the adequacy of the process to control other physicochemical aspects should also be addressed in a validation report. Depending upon the particular substance, these include potency, impurities, particulate matter, particle size, solvent residues, moisture content, and blend uniformity. For example, if the bulk substance is a blend of two active substances or an active substance and excipient, then there should be some discussion/evaluation of the process for assuring uniformity. The process validation report for such a blend would include documentation for the evaluation and assurance of uniformity. A list of validation reports and process variables evaluated should be reviewed.

As with a non-sterile bulk drug substance, there should be an impurity profile and specific, validated analytical methods. Those should also be reviewed.

Manufacturers are expected to validate the aseptic processing of sterile BPCs. Such validation must encompass all parts, phases, steps, and activities of any process where components, fluid pathways, in-process fluids, etc., are expected to remain sterile. Furthermore, such validation must include all probable potentials for loss of sterility as a result of processing. Validation must also account for all potential avenues of microbial ingress associated with the routine use of the process.

The validation procedure should approximate as closely as possible all those processing steps, activities, conditions, and characteristics that may have a bearing on the possibility of microbial ingress into the system during routine production. In this regard, it is essential that validation runs are as representative as possible of routine production to ensure that the results obtained from validation are generalizable to routine production.

Validation must include the 100% assessment of sterility of an appropriate material that is subjected to the validation procedure. Culture media is the material of choice. whenever feasible. Where not feasible, non-media alternatives would be acceptable. Where necessary, different materials can be used in series for different phases of a composite aseptic process incapable of accommodating a single material. In any event, some material simulating the sterile BPC, or the sterile BPC itself, must pass through the entire system that is intended to be sterile. Any material used for process validation must be microbiologically inert.

Environmental and personnel monitoring must be performed during validation, in a manner and amount sufficient to establish appropriate monitoring limits for routine production.

At least three consecutive, successful validation runs are necessary before an aseptic process can be considered to be validated.

Alternative proposals for the validation of the aseptic processing of bulk pharmaceuticals will be considered by FDA on a case-by-case basis. For example, it may be acceptable to exclude from the aseptic processing validation procedure certain stages of the post-sterilization bulk process that take place in a totally closed system. Such closed systems should be sterilized in place by a validated procedure, integrity tested for each lot, and should not be subject to any intrusions whereby there may be the likelihood of microbial ingress. Suitable continuous system pressurization would be considered an appropriate means for ensuring system integrity.

Water For Injection

Although water may not be a component of the sterile drug substance, water that comes in contact with the equipment or that enters into the reaction can be a source of impurities (e.g., endotoxins). Therefore, only water for injection should be utilized.

Some manufacturers have attempted to utilize marginal systems, such as single pass Reverse Osmosis (RO) systems. For example, a foreign drug substance manufacturer was using a single pass RO system with post RO sterilizing filters to minimize microbiological contamination. This system was found to be unacceptable. RO filters are not absolute and should therefore be in series. Also, the use of sterilizing filters in a Water for Injection system to mask a microbiological (endotoxin) problem has also been unacceptable. As with environmental monitoring, periodic reports should be reviewed.

Terminal Sterilization

There are some manufacturers who sterilize bulk powders after processing, by the use of ethylene oxide or dry heat. Some sterile bulk powders can withstand the lengthy times and high temperatures necessary for dry heat sterilization. In the process validation for a dry heat cycle for a sterile powder, important aspects that should be reviewed include: heat penetration and heat distribution, times, temperatures, stability (in relation to the amount of heat received), and particulates.

With regard to ethylene oxide, a substantial part of the sterile bulk drug industry has discontinued the use of ethylene oxide as a "sterilizing" agent. Because of employee safety considerations, ethylene oxide residues in product and the inability to validate ethylene oxide sterilization, its use is on the decline. As a primary means of sterilization, its utilization is questionable because of lack of assurance of penetration into the crystal core of a sterile powder.

Ethylene oxide has also been utilized in the treatment of sterile powders. Its principal use has been for surface sterilization of powders as a precaution against potential microbiological contamination of the sterile powder during aseptic handling.

There are some manufacturers of ophthalmics that continue to use it as a sterilant for the drug used in the formulation of sterile ophthalmic ointments and suspensions. If used as a primary sterilant, validation data should be reviewed. Refer to the Inspection Guide for Topical Products for further discussion.

Rework and Reprocessing

As with the principal manufacturing process, reprocessing procedures should also be validated. Additionally, these procedures must be approved in filings.

Review reprocessed batches and data that were used to validate the process. Detailed investigation reports, including the description, cause, and corrective action should be available for the batch to be reprocessed.

Laboratory Testing and Specifications

The sterility testing of sterile bulk substances should be observed. Additionally, any examples of initial sterility test failures should be investigated. The release of a batch, particularly of a sterile bulk drug substance, which fails an initial sterility test and passes a retest is very difficult to justify. Refer to the Microbiological Guide and Laboratory Guide for additional direction.

Particulate matter is another major concern with sterile powders. Specifications for particulate matter should be tighter than the compendial limits established for sterile dosage forms. The subsequent handling, transfer and filling of sterile powders increases the level of particulates. It is also important to identify particulates so that their source can be determined. Review the firm's program for performing particulate matter testing. If there are no official limits established, review their release criteria for particulates, and the basis of their limit.

With regard to residues, since some sterile powders are crystallized out of organic solvents, low levels of these solvents may be unavoidable. In addition to evaluation of the process to assure that minimal levels are established, data used by the firm to establish a residue level should be reviewed. Obviously, each batch should be tested for conformance with the residue specification. Refer to the Inspection Guide for Bulk Drug Substances for additional direction regarding limits for impurities.

Packaging Materials

Sterile bulk drug substances are filled into different type containers which include sterile plastic bags and sterile cans. With regard to sterile bags, sterilization by irradiation is the method of choice because of the absence of residues. There are some manufacturers, particularly foreign, which utilize formaldehyde. A major disadvantage is that formaldehyde residues may and frequently do appear in the sterile drug substance. Consequently, we have reservations about the acceptability of the use of formaldehyde for, container sterilization because of the possibility of product contamination with formaldehyde residues.

If multiple sterile bags are used, operations should be performed in aseptic processing areas. Since the dosage form manufacturer expects all inner bags to be sterile, outer bags should be applied over the primary bag containing the sterile drug in an aseptic processing area. One large manufacturer of a sterile powder only applied the immediate or primary bag in an aseptic processing area. Thus, the outer portion of this primary bag was contaminated when the other bags were applied over this bag in non-sterile processing areas.

With regard to sterile cans, a concern is particulates, which can be generated due to banging and movement. Because of some with trace quantities of aluminum, companies have moved to stainless steel cans.

The firm's validation data for the packaging system should be reviewed. Important aspects of the sterile bag system include residues, pinholes, foreign matter (particulates), sterility and endotoxins. Important aspects of the rigid container systems include moisture, particulates and sterility.

CHAPTER 8

Computerised Systems Inspection Guide

Introduction

Computers and automated equipment are an integral part of both the medical device and pharmaceutical industries. No longer are computerized systems limited to large manufacturing companies, smaller indigenous companies and start-ups can now employ affordable technology to manufacture medical and medicinal products.

Computer systems are used across a wide variety of applications including, material resourcing and control, quarantine systems for drug components, control of significant steps in manufacturing control of laboratory functions, storage and processing of raw data, management of warehousing and distribution. The first step when an auditor or inspector encounters a computer system is to understand the intended use of the system and gain a broad overview of it. Understanding the processes and functions that are under computer control or monitoring helps to shape the questions and scope of inspection.

Hardware

Each computerized system should have a corresponding schematic drawing of the hardware. The drawing does not need to have every technical detail, and should simply include an overview of the major input devices, output devices, signal converters, central processing unit, distribution systems, significant peripheral devices and how they communicate and how they are linked.

Input Devices: Equipment which translates external information into electrical pulses which the computer can understand. Examples are thermocouples, flow meters, load cells, pH meters, pressure gauges, control panels, and operator keyboards.

Examples of functions are:

➢ Operator keyboard used to enter batch data

➢ A thermocouple that provides temperature input for sterilization calculations

➢ Flow meter provides volume of liquid component going into a mixing tank

Output Devices: Equipment which receives electrical pulses from the computer and either causes an action to occur, generally in controlling the manufacturing process, or passively records data.

Examples include:

➢ printers
➢ alarms
➢ motors

➤ solenoids
➤ HMIs
➤ valves
➤ switches

Examples of functions are:

➤ Solenoid activates the impeller of a mixer
➤ Valve controls the amount of steam delivered to a sterilizer
➤ Printer records significant events during sterilization process
➤ Signal Converters

Peripheral Devices: all computer associated devices external to the CPU can be considered peripheral devices. This includes the previously discussed input and output devices. Many peripheral devices can be both input and output, they are commonly known as I/O devices such as printers and keyboards.

Validation of Hardware

The suitability of computer hardware for the tasks assigned to pharmaceutical production must be demonstrated through appropriate tests and challenges. The depth and scope of hardware validation will depend upon the complexity of the system and its potential affect on drug quality.

The validation program need not be elaborate but should be sufficient to support a high degree of confidence that the system will consistently do what it is supposed to do. In considering hardware validation the following points should be addressed:

Does the capacity of the hardware match its assigned function? For example, in a firm using a computer system to maintain its labeling text, including foreign language labeling, do the CRT and printer have the capacity to write foreign language accent marks?

Have operational limits been identified and considered in establishing production procedures? For example, a computer's memory and connector input ports may limit the number of thermocouples a computer can monitor. These limits should be identified in the firm's standard operating procedures.

Have test conditions simulated "worst case" production conditions? A computer may function well under minimal production stress (as in vendor's controlled environment) but falter under high stresses of equipment speed, data input overload or frequent or continuous multi-shift use (and a harsh environment). Therefore, it is insufficient to test computer hardware for proper operation during a one hour interval, when the system will be called upon in worst case conditions to run continuously for 14 days at a time. Some firms may test the circuits of a computer by "feeding" it electrical signals from a signal simulator. The simulator sends out voltages which are designed to correspond to voltages normally transmitted by input devices. When simulators are connected to the computer, the program should be executed as if the emulated input devices were actually connected. These signal simulators can be useful tools for validation; however, they may not pose worse case conditions and their accuracy in mimicking input device performance should be established. In addition, validation runs should be accomplished on line using actual input devices. Signal simulators can also be used to train employees on computer operations without actually using production equipment.

Have hardware tests been repeated enough times to assure a reasonable measure of reproducibility and consistency? In general, at least three test runs should be made to cover different operating conditions. If test results are widely divergent they may indicate an out of control state.

Has the validation program been thoroughly documented? Documentation should include a validation protocol and test results which are specific and meaningful in relation to the attribute being tested. For example, if a printer's reliability is being tested it would be insuess the results merely as "passes," in the absence of other qualifying data such as printing speeds, duration of printing, and the number of input feeds to the printing devices.

Are systems in place to initiate revalidation when significant changes are made? Revalidation is indicated, for example, when a major piece of equipment such a circuit board or an entire CPU is replaced. In some instances identical hardware replacements may adequately be tested by the use of diagnostic programs available from the vendor. In other cases, as when different models of hardware are introduced, more extensive testing under worst case production conditions, is indicated.

Much of the hardware validation may be performed by the computer vendor. However, the ultimate responsibility for suitability of equipment used in drug processing rests with the pharmaceutical manufacturer. Hardware validation data and protocols should be kept at the drug manufacturer's facility. When validation information is produced by an outside firm, such as the computer vendor, the records maintained by the drug establishment need not be all inclusive of voluminous test data; however, such records should be reasonably complete (including general results and protocols) to allow the drug manufacturer to assess the adequacy of the validation. A mere certification of suitability from the vendor, for example, is inadequate.

Software

Software is the term used to describe the total set of programs used by a computer. These programs exist at different language levels, generally the higher the level, the closer the text is to human language. These levels are set forth below. During the inspection identify key computer programs used by the firm. Of particular importance are those programs which control and document dosage form production and laboratory testing. Usually a firm can readily list the names of such programs on a CRT display or in hard copy. Such a list is sometimes called a menu or main menu.

Levels

Machine Language: This contains coded instructions, represented by binary numbers. The computer executes this code.

High Level Language: This language is characterized by a vocabulary of English words and mathematical symbols. These are source programs which must be translated by a compiler or interpreter into an object program.

Application Language: This is generally based on a high level language but modified for a specific industry application and uses the vocabulary of that industry.

Fixed Setpoint. This is the desired value of a process variable which cannot be changed by the operator during execution. Determine major fixed setpoints, such as desired time/temperature curve, desired pH, etc. Time may also be used as a set point to stop the process to allow the operator to interact with the processing.

Variable Set point. This is the desired value of a process variable which may change from run to run and must usually be entered by the operator. For example, entering one of several sterilizer load patterns into a sterilization computer process.

Edits. A program may be written in such a manner as to reject or alter certain input or output information which does not conform to some pre-determined criterion or otherwise fall within certain pre-established limits. This is an edit and it can be a useful way of minimizing errors; for example, if a certain piece of input data must consist of a four character number, program edits can be used to reject erroneous entry of a five character number or four characters comprised of both numbers and letters. On the other hand, edits can also be used to falsify information and give the erroneous impression that a process is under control; for example, a program output edit may add a spurious "correction" factor to F values which fall outside of the pre-established limits, thus turning an unacceptable value into an ue. It is, therefore, important to attempt to identify such significant program edits during the inspection, whenever possible.

Sometimes such edits can manifest themselves in unusually consistent input/output information.

Input manipulation.

Key Points

Software Development:

During the inspection determine if the computer programs used by the firm have been purchased as "canned" from outside vendors, developed within the firm, prepared on a customized basis by a software producer, or some combination of these sourams are highly specialized and may be licensed to pharmaceutical establishments. If the programs used by the firm are purchased or developed by outside vendors determine which firms prepared the programs.

In some cases "canned" or customized programs may contain segments (such as complex algorithms) which are proprietary to their authors and which cannot normally be readily retrieved in program code without executing complex code breaking schemes. In these cases the buyer must accept on faith that the software will perform properly. If the drug manufacturer is using such a program to control or monitor a significant process, determine what steps the firm has taken to assure itself that such program blind spots do not compromise the program performance.

Where drug firms develop their own application programs, review the firm's documentation of the approval process. This approval process should be addressed in the firm's SOP. It may be useful to review the firm's source (English) documents which formed the basis of the programs.

Software Security

Determine how the firm prevents unauthorized program changes and how data are secure from alteration, inadvertent erasures, or loss (21 CFR 211.68). Some computers can only be operated in a programming mode when two keys are used to unlock an appropriate device. When this security method is used, determine how use of keys is restricted. Another way of achieving program security is the use of ROM (read only memory), PROM (programmable read only memory), or EPROM (erasable programmable read only memory) modules within the computer to "permanently" store programs. Usually, specialized equipment separate from the computer is needed to change an EPROM or establish a program in PROM so that changes would not be made by the operator.

A program in EPROM is erase the module (which has a quartz window) to ultraviolet light. In these cases a program is secure to the extent it can't be over-ridden by the operator. Determine who in the firm has the ability and/or is authorized to write, alter or have access to programs. The firm's security procedures should be in writing. Security should also extend to devices used to store programs, such as tapes, disks and magnetic strip cards. Determine if accountability is maintained for these devices and if access to them is limited. For instance, magnetic strip cards containing a program to run a sterilization cycle may be kept in a locked cabinet and issued to operators on a charge-out basis with return of the card immediately after it is used.

Validation of Software

It is vital that a firm have assurance that computer programs, especially those that control manufacturing processing, will consistently perform as they are supposed to within pre-established operational limits. Determine who conducted software validation and how key programs were tested. In considering software validation the following points should be addressed:

Does the program match the assigned operational function? For example, if a program is assigned to generate batch records then it should account for the maximum number of different lots of each component that might be used in the formulation. Consider what might happen when three lots of a component are used with a program designed to record lot designations and quantities for up to two different lots of each component.

The first lot may be accurately recorded, but the next two lots might be recorded as a single quantity having the second lot designation; the resultant computer generated record therefore would fail to show the use of three different lots and the quantities of each of the second and third lots going into the mixture.

Have test conditions simulated "worst case" production limits? A program should be tested, for example, under the most challenging conditions of process speed, data volume and frequency. Date should be considered in this aspect of validation. For example, the number of characters allowed for a lot number should be large enough to accommodate the longest lot number system that will be used.

Have tests been repeated enough times to assure consistent reliable results? Divergent results from replicate data entries may signify a program bug. In general, at least three separate runs should be made.

Has the software validation been thoroughly documented? Documentation should include a testing protocol and test results which are meaningful and specific to the attribute being tested; individuals who reviewed and approved the validation should be identified in the documentation.

Are systems in place to initiate revalidation when program changes are made? If process parameters such as time/temperature, sequence of program steps, or data editing/handling are changed then revalidation is indicated. Although much of the software validation may be accomplished by outside firms, such as computer or software vendors, the ultimate responsibility for program suitability rests with the pharmaceutical manufacturer.

Records of software validation should be maintained by the drug establishment, although when conducted by outside experts such records need not be voluminous but rather complete enough (including protocols and general results) to allow the drug manufacturer to assess the adequacy of the validation. Mere vendor certification of software suitability is inadequate. Signal simulators many be used in software validation. These are discussed in point No. 3 of Validation of Hardware.

CGMP Guidance

Hardware

Computer hardware is classified as equipment. If the equipment has the potential to impact quality then it falls under CGMP regulations.

The following section of 21 CFR applies to equipment hardware.

➢ 21 CFR 211.63 repment be suitably located to facilitate operations for the equipment's intended use

➢ 21 CFR 211.67 requires a maintenance program for equipment

➢ 21 CFR 211.68(a) states that computers may be used and requires a calibration program

Software

In general, software is regarded as records or standard operating procedures (instructions) within the meaning of the CGMP regulations and the corresponding sections of the CGMP regulations apply, for example:

Record Controls. 21 CFR 211.68(b) requires programs to ensure accuracy and security of computer inputs, outputs, and data.

Record Access. 21 CFR 211.180(c) states that records required by the regulations shall be available as part of an authorized inspection at the establishment for inspection and are subject to reproduction. Computer records retrievable from a remote location are acceptable

In considering the copying of electronic records however, the act of copying must be reasonable, as the word reasonable is used in the FD& C Act to limit how we may conduct inspections. In some cases it may be reasonable to copy a disk or tape whereas in other cases it might not, particularly where we would have to physically remove the disk or tape from the establishment in order to copy it. (Consider the analogy of removing an entire file cabinet so that we can copy five pieces of paper.) We believe that, rather than copy an entire disk or tape ourselves, it is preferable to have the firm generate hard copies of only those portions of the disk or tape which we need to document.

Record Medium. 21 CFR 211.180(d) states that retained records may be originals or true copies and, when necessary, ocopying equipment shall be available. This concept applies to magnetic tape and disks.

Record Retention. 21 CFR 211.180(a) states record retention requirements. They are the same for electronic media and paper.

Computer Programs. FD& C Act. Section 704(a), for prescription drug products, would allow inspectional access to computer programs if such inspection is performed within the constraint of being reasonable.

There are several factors which must be considered on a case by case basis in determining what is reasonable in accessing a firm's computer. For example, the effect on drug production is a factor; specifically, if the process of running a program disrupts drug production in an adverse manner then that would be unreasonable. Another factor is whether or not our manipulations give us access to unauthorized information; the data we may be searching with a program may contain some information we are not entitled to review such as financial data. Consider also that some computer programs are protected by copyright and carefully licensed to software users; thus, we would not be able to copy and use such programs without prior approval of their owners.

Record Review. 21 CFR 211.180(e) states that where appropriate records associated with every batch shall be reviewed as part of a periodic review of quality standards. It is acceptable for a firm to conduct part of the review by running a computer program which culls out analytical data from each batch and conducts trend analysis to determine the need to change product specifications, manufacturing methods, or control prata itself must be meaningful (i.e., specified and relevant to enable an evaluation to be performed). It is not necessary to review each and every bit of information on the batch record. However, the computerized trend analysis data would constitute only a portion of the data which must be reviewed. A review must also be made of records of complaints, recalls, returned or salvaged products, and investigations of unexpected production discrepancies (e.g., yield reconciliations) and any failures of batches to meet their specifications. This information is usually separate from conventional batch records and so would not necessarily be reviewed by the trend analysis program.

QC Record Review. 21 CFR 211.192 requires the quality control unit to review and approve production and control records prior to batch release/distribution. If this record screening review (to check errors and anomalies) is computerized and is at least as comprehensive and accurate as a manual review, then it is acceptable for the QC unit to review a computer generated exception report as part of the batch release. The batch record information required by the regulation must still be retained. It is also important that the accuracy and reliability of the screening program be demonstrated. It is unlikely however, that all production and control records will be computerized; labeling, packaging, and analytical records may still be in manual form and would therefore be manually reviewed.

Documentation

21 CFR 211.188(b) (11) requires that batch production and control records include identification of each person who conducts, supervises or checks each significant step in the process. The intent is to assure that each step was, in fact, performed and that there is some record to show this, from which the history of the lot could be traced. It is quite possible that an automated system can achieve the same, or higher, level of assurance in which case it may not be necessary to have persons document the performance of each event in a series of unbranched automated events on the production line. For example, let us say an automated/computerized system is designed to perform steps

Electronic Records & Signatures

Part 11 of the FDA CFR is relevant to "records in electronic form that are created, modified, maintained, archived, retrieved, or transmitted under any records requirements set forth in Agency regulations." This first section of the book provides a background information and explanations of each section and requirement of the regulation. The second half of this eBook provides a clear and transferrable verification process for each requirement of 21 CFR Part 11, with suggested verification methods included.

As of 2007, several sections of the regulation have been identified as excessive and the FDA announced in guidance that it will exercise enforcement discretion on some parts of 21 CFR part 11. This has been welcomed by some manufactures but it has also causes a degree of confusion.

The requirements relating to access controls are the most fundamental requirements and are routinely enforced. The "predicate rules" that required organizations to keep records the first place are still in effect. If electronic records are illegible, inaccessible, or corrupted, manufacturers are still subject to those requirements.

If a regulated firm keeps "hard copies" of all required records, those paper documents can be considered the authoritative document for regulatory purposes. This then means that the computer system is not in scope for electronic records requirements, although subject to predicate rules which still require validation.

If the "hard copy" is to be identified as the authoritative document, the "hard copy" must be a complete and accurate copy of the electronic source. The manufacturer must use the hard copy (rather than electronic versions stored in the system) of the records for regulated activities.

Definition of Records

The FDA has deemed the following records or signatures in electronic format subject to 21 CFR part 11:

"Records that are required to be maintained under predicate rule requirements and that are maintained in electronic format in place of paper format. On the other hand, records (and any associated signatures) that are not required to be retained under predicate rules, but that are nonetheless maintained in electronic format, are not part 11 records.

Records that are required to be maintained under predicate rules, that are maintained in electronic format in addition to paper format, and that are relied on to perform regulated activities. Records submitted to FDA, under predicate rules (even if such records are not specifically identified in Agency regulations) in electronic format (assuming the records have been identified in docket number 92S-0251 as the types of submissions the Agency accepts in electronic format). However, a record that is not itself submitted, but is used Contains Nonbinding Recommendations in generating a submission, is not a part 11 record unless it is otherwise required to be 205 maintained under a predicate rule and it is maintained in electronic format.

Electronic signatures that are intended to be the equivalent of handwritten signatures, initials, and other general signings required by predicate rules. Part 11 signatures include electronic signatures that are used, for example, to document the fact that certain events or actions occurred in accordance with the predicate rule (e.g. approved, reviewed, and verified)."

The above definitions are taken from FDA guidance document entitled "FDA Guidance for Industry: 21 CFR Part 11 - Electronic Records and Electronic Signatures: Scope and Application, August 2003." This document also provides recommendations on documenting key decisions that may be taken in relation to 21 CFR Part 11 applicability and compliance.

Requirements and Specifications

The need for compliance to 21 CFR depends on type of technology and level of automation and computerisation involved in the manufacturing process or other actives that are GxP impacting. Does the system store electronic records? Does the system require a login? Is there an audit trial? If a complex system is to be procured, the requirements need to be communicated to the manufacturer as part of a User requirement specification and/or software requirement specification.

General Guidance on Requirement Specifications

While the Quality System regulation states that design input requirements must be documented, and that specified requirements must be verified, the regulation does not further clarify the distinction between the terms "requirement" and "specification." A requirement can be any need or expectation for a system or for its software. Requirements reflect the stated or implied needs of the customer, and may be market-based, contractual, or statutory, as well as an organization's internal requirements.

There can be many different kinds of requirements (e.g., design, functional, implementation, interface, performance, or physical requirements). Software requirements are typically derived from the system requirements for those aspects of system functionality that have been allocated to software. Software requirements are typically stated in functional terms and are defined, refined, and updated as a development project progresses. Success in accurately and completely documenting software requirements is a crucial factor in successful validation of the resulting software. Page 6 Guidance for Industry and FDA Staff General Principles of Software Validation A specification is defined as "a document that states requirements." (21 CFR 820.3(y)) It may refer to or include drawings, patterns, or other relevant documents and usually indicates the means and the criteria whereby conformity with the requirement can be checked.

There are many different kinds of written specifications, e.g., system requirements specification, software requirements specification, software design specification, software test specification, software integration specification, etc. All of these documents establish "specified requirements" and are design outputs for which various forms of verification are necessary.

Validation of Computerised Systems

The requirement for computerised systems to be compliant to 21 CFR part 11, needs to be identified early on the project to ensure that the vendor or supplier of the systems or equipment can develop, build a system that meets the requirements of 21 CFR part 11. Computer system validation can be divided into 3 distinct phases which include: (1) Plan, (2) Design & Development, (3) verification and (4) Retirement. The requirement for computerised systems to be compliant to 21 CFR part 11, needs to be identified early on the project to ensure that the vendor or supplier of the systems or equipment can develop, build a system that meets the requirements of 21 CFR part 11.

Plan: This phase involves the planning of the validation effort required to implement the system and identification of key milestones and requirements. It requires supplier assessments, assessments of the regulatory and system risks, supplier, development of a validation approach and the identification of deliverables that will be generated, that will support the implementation and operation of the system.

Design & Development: This phase consists of the design, development and configuration of the hardware and software required to meet the system requirements. In case of custom software, design and developmental testing is important to ensure proper functionality prior to verification testing.

Verification: This phase confirms that requirements and specifications have been met. Testing is required to ensure the system operates as intended. Upon successful testing and verification, the system can be released for use. Once verification activities have begun any changes to the system must managed through change control. In case of successful completion of the verification activities (i.e. any deviation has been evaluated and addressed), the system is released for effective use. Operation This phase supports the need to maintain compliance and fitness for intended use after the system is accepted and released for use.

Retirement: This phase consists of the planning, executing and summarizing of the events required for system shutdown. It includes the appropriate handling of the supporting documents and the data contained within the system. While described here as a separate phase, a system's retirement can be handled as part of a new system implementation or as a separate project.

Best practice when it comes to Computer System validation is to adopt a life cycle approach for computer systems which requires the completed of activities in a systematic way from system conception to retirement. Life cycle activities could be scaled according to system impact on product quality, patient safety and data integrity, system complexity and novelty, supplier assessment and business risk.

Computer System: A computer / automated system consisting of the hardware, software, and network components, together with the controlled functions (personnel, procedures, and equipment) and associated documentation.

Computer System Validation: A process that confirms by examination and provision of objective evidence that the computer system conforms to user needs and intended uses. Computer System validation is a process for achieving and maintaining compliance with GxP regulations and fitness for intended use by adoption of life cycle activities, deliverables, and controls.

GxP Regulated Computer Systems: Computer systems determined to have a potential impact on Product Quality, Patient Safety and Data Integrity; these systems are required to comply with the relevant GxP regulations.

Data Integrity: is the degree to which data is reliable and without error. Data must be accurate, attributable, contemporaneous, original, legible and available. A breach of data integrity occurs when any person manipulates or distorts data and submits the results of that data as valid.

Predicate rules: a predicate rule is any FDA regulation that requires companies to maintain certain records and submit information to the agency as part of compliance.

To gain a better understanding of the validation of computerized systems, consult the following publication- "FDA's guidance for industry and FDA staff General Principles of Software Validation." Industry guidance such as the GAMP 5 guide issued by ISPE is also a useful reference.

Electronic Records

When it comes to the regulated industries such as the medical device industry, every process and procedure must be documented. Documentation ensures that everyone is working in the same manner with the same procedures. However, documentation is more than just writing down procedures and processes. It is also concerned with how documents are controlled, how they are updated and how they are stored.

Electronic Document management systems

Electronic document management systems aka EDMS are now the norm and gold standard for most medium to large organisations. Many companies that provide medical device manufacturers with an EDMS can be customised to match the business processes particular to an organisation. With configurable or customisable software, validation and proper verification is important to ensure the system operates as intended. There are also regulatory requirements that stipulate the expectations and requirements of such system. For example, the application of electronic signatures and the presence of audit trials. FDA 21 CFR Part 11 details the requirements with regards to electronic records and electronic signatures. For medicinal products in Europe, GMP V4 Annex 11 specifies similar requirements.

Record Retention

Regard to the part 11 requirements for the protection of records to enable their accurate and ready retrieval throughout the records retention period (11.10 (c)) Persons must also comply with all applicable predicate rule requirements for record retention and availability such as (211.180(c) general requirements. The decision to follow 21 CFR part 11 should be justified and documented as part of a risk assessment and based on the value of the records over time.

FDA does not object to archiving of required records in electronic format to non-electronic media such as paper, or to a standard electronic file format (examples of such formats include, but are not limited to, PDF, XML, or SGML). Persons must still comply with all predicate rule requirements, and the records themselves and any copies of the required records should preserve their content and meaning. As long as predicate rule requirements are fully satisfied and the content and meaning of the records are preserved and archived, you can delete the electronic version of the records. In addition, paper and electronic record and signature components can co-exist as long as predicate rule requirements are met and the content and meaning of those records are preserved.

Electronic Signatures

Electronic signatures are computer-generated character strings that count as the legal equivalent of a handwritten signature. The regulations for the use of electronic signatures are set out in 21 CFR Part 11 of the FDA. Each electronic signature must be assigned uniquely to one person and must not be used by any other person. It must be possible to confirm to the authorities that an electronic signature represents the legal equivalent of a handwritten signature. Electronic signatures can be biometrically based or the system can be set up without biometric features.

Conventional Electronic Signatures

If electronic signatures are used that are not based on biometrics, they must be created so that persons executing signatures must identify themselves using at least two identifying components. This also applies in all cases in which a chip card replaces one of the two identification components. These identifying components, can, for example consist of a user identifier and a password. The identification components must be assigned uniquely and must only be used by the actual owner of the signature.

When owners of signatures want to use their electronic signatures, they must identify themselves by means of at least two identification components. The exception to this rule is when the owner executes several electronic signatures during one uninterrupted session. In this case, persons executing signatures need to identify themselves with both identification components only when applying the first signature. For the second and subsequent signatures, one unique identification component (password) is then adequate identification.

Audit Trail

Title 21 CFR details predicate rule requirements relating to documentation of, for example, date time, or the sequencing of events, as well as any requirements for ensuring that changes to records do obscure previous entries.

Making the decision on whether to apply audit trails, or other appropriate measures, or on the need to comply with predicate rule requirements should involve a justified and documented risk assessment. Any Risk assessment should determine the potential effect on product quality and safety and the integrity of the record.

Change Management

Validation programs are subject to change control. Each company or organisation should have a procedure detailing the change management process. Below is a suggested overview of a typical change control process.

Any system, facility, document or process that has the potential to impact product quality and validated state is generally subject to following a change control process.

Another term used in industry is Enterprise Change Control or Engineering Change Control. Essentially these terms are the same. The intent is to control and manage change consistently.

A change control can take the form of a document which drives the agenda and the specific requirement. Change control is also created with enterprise software such as Kintana, Documentum and SAP. While each company will have varying processes, some basics are common. These include the 3 stages of change control; pre-implementation, implementation and post implementation (if required). Below, 2 case studies are detailed where there is a change in manufacturing which requires a formal change control process to be applied.

Summary of 21 CFR Part 11.10

11.10 (a) Accuracy, reliability & consistent intended performance.

11.10 (b) Copies of records (Paper)- complete copies of records in both human readable and electronic form suitable for inspection, review, and copying.

11.10 (b) Copies of records (Electronic)- complete copies of records in both human readable and electronic form suitable for inspection, review, and copying.

11.10 (c) Protection of records- accurate and allow ready-retrieval throughout the records retention period.

11.10 (d) & (g) Authorised access - Limiting system access to authorized individuals in relation to accessing records, the operation or computer system input or output devices, altering of records, or performing the operation at hand.

11.10 (e) Computer systems (including hardware and software), controls, and attendant documentation maintained under this part shall be readily available for, and subject to, FDA inspection.

11.10 (f) Sequencing of steps - checks to enforce permitted sequencing of steps and events

11.10 (h) Input device authorisation - checks the validity of the source of data.

11.10 (i) Input device persons education, training, and experience

11.10 (j) Establishment of written policies - establishment of, and adherence to, written policies that hold individuals accountable and responsible for actions initiated under their electronic signatures, in order to deter record and signature falsification.

11.10 (k) Appropriate Controls - appropriate controls over systems documentation-(1) Adequate controls over the distribution of, access to, and use of documentation for system operation and maintenance. (2) Revision and change control procedures to maintain an audit trail that documents time-sequenced development and modification of systems documentation.

Electronic Records Verification Methods

(21 CFR 11.10 Electronic Records)

This section provides some simple verification methods for electronic records

11.10 (a) Accuracy, reliability & consistent intended performance

Verification Method: create test scripts to verify that all types of records generated and maintained by the system are accurate, consistent and contain the intended data. Repeat tests using different challenge conditions to cover any anticipated operating conditions. Ensure that test scripts contain the relevant acceptance criteria.

11.10 (b) Copies of Paper based records

Verification Method: Create paper copies for the record types under test and verification. Compare the paper hardcopies with electronic records that are stored and displayed in the computer system.

11.10 (b) Copies of electronic records

Verification Method: Create electronic copies (softcopies) and compare the copy to that contained in the computer system.

11.10 (c) Protection of records

Verification Method: Review the controls and procedures that ensure accuracy and ready retrieval throughout the records retention period.

11.10 (d) & (g) Authorised access
Verification Method: Complete security verification for each account type e.g. operator manager etc. Verify that access is granted and denied via the login function.

11.10 (e) shall be readily available for, and subject to, FDA inspection
Verification method: maintain periodic inspections of the system

11.10 (f) Sequencing of steps

Verification Method: The purpose of this requirement is to ensure the correct sequencing of steps and events occurs. Create tests scripts to demonstrate that any series of steps are operating as intended e.g. approval or signature or login.

Note: This requirement relates to electronic records and is not intended to verify any Operational or functional sequences. These are typically covered in Equipment Validation.

11.10 (h) Input device authorisation
Verification method: The purpose of this step is to check the validity of the source of data. When data is networked or transferred form a device the user must verify that the device is a valid input device. This can be achieved by code review or physical testing.

11.10 (i) Input device person's education, training, and experience

Verification method: Ensure persons using the system are appropriately educated, trained and experienced. Attach evidence of same.

11.10 (j) Establishment of written policies
Verification method: verify the availability of written policies in relation to responsibilities for actions taken under their electronic signatures.

11.10 (k) Appropriate Controls - appropriate controls over systems documentation.

Verification method: review access to SOPs to verify their distribution and use for operation of the system is controlled. Verify that chance control is practiced for system related documentation.

Data Integrity

Data generated by or used in GxP impacting activities must be handled and protected in accordance with international and national regulatory requirements. The application of Data Integrity applies to many industries and products that touch the lives of patients and end users across the globe. Some examples of products that must meet Data Integrity regulations include (1) active pharmaceutical ingredients, (2) medical devices, (3) medicinal products, (4) vaccines and (5) cosmetics.

The below agencies and regulatory authorities provide specific requirements on Data Integrity:

➤ EU GMP – EudraLex – Rules governing medicinal products in the European Union Volume 4 – Guidelines to Good Manufacturing Practice for medicinal products for human use – Products for Human and Veterinary Use, Annex 11: Computerised Systems – (1, 7.2, 17)

➤ FDA – 21 CFR Part 11 – Food and Drug Administration – Electronic Records; Electronic Signatures – Scope and Application (C)

➤ FDA- 21 CFR Part 211 – Food and Drug Administration – Code of Federal Regulations -

Good Manufacturing Practices - 211.188a, 211.194.2, 211.194.8

➢ ICH E6 – International Conference on Harmonization - Guideline for Good Clinical Practice (5.2.1, 8.1, 8.3)

➢ MHRA – United Kingdom - Medicines and Healthcare Products Regulatory Agency - GMP
Data Integrity Definitions and Guidance for Industry (2015)

➢ PIC/S Guidance PI 011-] – Pharmaceutical Inspection Convention Scheme - Good
Practices for Computerised Systems in Regulated "GXP" Environments

Introduction

Within the Life science industry the saying goes "if it's not written down, it didn't happen". This is a powerful message that is a suitable starting point for Data integrity. In the current and present day, the mention of Data Integrity quickly conjures an image of excel sheets, big data, databases and computers in our minds. However, it has a broader impact with its roots in the basics of good science – good documentation.

Data integrity indeed does apply to "soft" or electronic data but also applies to paper based systems and records. GxP is the umbrella acronym that stands of "Good Practices" in all our tasks and activities, be it laboratory testing, process engineering and so on. A core element in meeting GxP is abiding to "Good Documentation Practices" (GDP). Having good written records is fundamental to patient and product safety with pharmaceutical, biopharmaceutical and medical devices industries. So, Data Integrity begins with the small stuff -Real-time data collection, real-time review, honest and accurate recording of data and events.

The Integrity of data relies on several factors. It can be influenced a company's culture or approach to doing business. It can also be affected by the level of experience of knowledge within a company. Many traditional engineering companies outside the regulated life science community simply do not have the need to be so thorough in their handling of data and information.

Within GxP environment, controls, training and the design and operation of systems and processes influence Data Integrity on a day to day basis. Most of the time, those affected by the controls or systems do not think of them, but they can either support or inhibit data integrity and the reliability of data. Obviously, equipment, systems and processes should be a key part in making Data reliable and accurate.

Key Terms

Configuration Identification

Software and hardware packages should be identified by a unique product identifier and a version number. For the software end-user, the parts of an automated system that are subject to configuration management should be clearly identified. The system should therefore be broken down into configuration items. These should be identified at an early phase of development so that a complete list of configuration items is defined and maintained. The application-specific items should have a unique name or version ID. The depth of detail when specifying the elements is decided by the needs of the system, and the organization developing that system.

Requirements for the User ID and Password

User ID: The user ID of a system should have a minimum length agreed with the customer and should be unique within the system.

Password: A password should always consist of a combination of numeric and alphanumeric characters. When setting up passwords, the number of characters and a period after which a password expires should be stipulated. The structure of the password is normally selected to suit the specific customer. The configuration is described in the section Security Settings of Password Policy.

Criteria for the structure of a password are as follows:

> ➢ Minimum length of the password
> ➢ Use of numeric and alphanumeric characters
> ➢ Case sensitivity

Audit Trail

The audit trail is a control mechanism of a system that allows all data entered or modified to be traced back to the original data. A reliable and secure audit trail is particularly important in conjunction with the creation, change or deletion of GMP relevant electronic records. In this case, the audit trail must archive and document all the changes or actions made along with the date and time. Typical contents of an audit trail must be recorded and describe the procedures "who changed what and when" (old value/new value).

Data: any data (numerical or otherwise) which is collected or processed as part of GxP activities in order to generate GxP documents and records using a paper-based or electronic process.

Data handling: any GxP task that involves creation, entry, review, approval, analysis, reporting, storage, archival, retrieval, or disposal of GxP data

Data integrity: degree to which a collection of GxP data is managed through effective organizational, operational, and technical mechanisms to ensure GxP data reliability.

Data lifecycle: starts from the time of data creation to the point of use and during its retention, archival, retrieval, and eventual disposal

GxP impacting: any action that can impact the quality or safety of a product or critical process.

Application: Software installed on a defined platform/hardware providing specific functionality.

Bespoke/Customized computerised system: A computerised system individually designed to suit a specific business process.

Commercial of the shelf software: Software commercially available, whose fitness for use is demonstrated by a broad spectrum of users.

IT Infrastructure: The hardware and software such as networking software and operation systems, which makes it possible for the application to function.

Life cycle: All phases in the life of the system from initial requirements until retirement including design, specification, programming, testing, installation, operation, and maintenance.

Process owner: The person responsible for the business process.

System owner: The person responsible for the availability, and maintenance of a computerised system and for the security of the data residing on that system.

Third Party: Parties not directly managed by the holder of the manufacturing and/or import authorisation.

The Lifecycle of Data

Regulations that speak to GxP and Data Integrity can apply to many different streams with the life science sector as previously mentioned. From medical devices to pharmaceuticals, all act in different manners, with long and short term applications. Take the example of a Total Knee Replacement. Many designs now ensure their effectiveness in excess of 10 years, even up to 20 years depending on individual circumstances. This requires many key records within manufacturing to be kept for several decades. Thus, data retention requirements specify the retention periods of such documents. The Integrity of GxP data must be protected during the entire data lifecycle. From creation of the data and records to the eventual destruction of data after the retention period is fulfilled.

Data integrity does not only apply to products it also applies to:

- ➢ Equipment
- ➢ Computerised Systems
- ➢ Test records
- ➢ Inspection records
- ➢ Material certificates

Data integrity ensures patient safety, product quality, and product supplies are generated by the product lifecycle processes. As such, the opportunities

Process Design

Failure to maintain data integrity can occur throughout the lifecycle of data; however, a thoughtful design of systems can prevent breaches in data and restrict the severity of any attempts to alter data. Therefore design, should aim to include controls and preventative measures. At a high level, this can be achieved by:

- ➢ Limiting access to GxP events and data
- ➢ Standard Operating Procedures (SOPs)
- ➢ Training
- ➢ System Owners

Data Reliability

Data reliability is the foundation to achieving cGxP data integrity. The FDA's ALOCA model can be used to enforce data reliability.

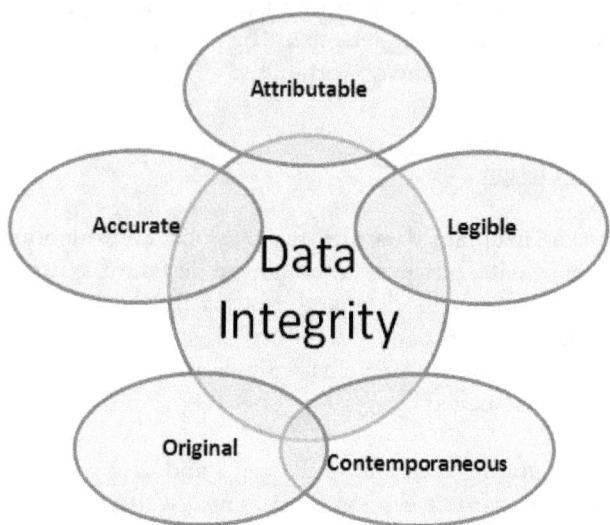

Figure: ALOCA graphical representation

Accuracy: the GxP data is recorded, calculated, analysed, and reported as found and correctly.

Attributable: any actions or calculations performance on GxP data can be attributed to or traceable to the person that performed the actions and the date and time at which they were performed.

Legible: the GxP data is recorded in a clear and human readable form

Contemporaneity: the GxP data is recorded at the same time as the observation/measurement is made or as soon as possible after the event.

Original: the initial data recorded is available and not altered.

An additional point to make it that of trustworthiness. It is assumed that engineers and scientists etc. working across the Life science industries are ethical and do not falsify data or information. Typically companies can implement a code or practice or Ethical behaviour program to desist people from intentional unethical behaviour or the falsification of records.

The Journey of Data

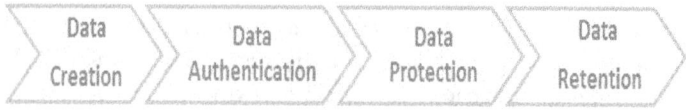

Figure: Data Lifecycle

Data Creation: the point at which the values or data is created. The data and information is original (raw).

Data Authentication: Within a GxP environment, authentication refers to the approval of data (electronic signatures).E-signatures are key controls within software that prompt the user to enter a unique username and password to acknowledge a recording or action. The E-signature should create a permanent link with the electronic record that cannot be removed and can be viewed through an audit trial.

Data Protection: Once the data is created, the handling of the data must ensure data integrity. For electronic data, this includes access control to computer systems. Other practical restrictions can also be made such as limiting room and site access to authorised personnel.

Data Retention: The controlled storage, backup and arching of data. Retention of records may be required for several decades depending on the type of data and the regulatory requirements relating the particular product or industry.

Technical Controls

The benefits of modern software and computerised systems allow robust and complex data handling and calculations to be completed. With this modern capability that is becoming more powerful, comes more responsibility with regard to the use of data.

The Computerised systems used to generate, gather or interpret GxP data must be fulfil several criteria. First and foremost they must be fit for the intended use. The software and hardware must be validated and proven to be consistent and reliable. Some general considerations for the use of Computerised systems include:

> Design to foster integrity of GxP data
> User requirements specification detailing the intended use and required functionality
> An approved vendor with certification to ISO 9001 or other Quality management standards
> Software should meet the requirements of regulations such as FDA 21 CFR Part 11.
> Written procedures on how automated processes function.

It should not be an easy process from persons to alter or corrupt data when using computerised systems. GxP impacting Computer systems should have controls that prevent unauthorised access along with audit trail history.

Audit trail design and configuration capture key critical processes, events, settings and information. This enables any investigations of quality events impacting data integrity to be reviewed and analysed.

Practical Elements to Data Integrity

Facilities and systems must be configured in a way that encourages compliance with principles of data integrity. Examples include:

- ➢ Availability of clocks for recording times.
- ➢ Access points to allow swift reference to GxP records at locations where tasks are completed.
- ➢ Control of raw data.
- ➢ Control of approved documents.

Organisational Controls

Regulated companies such as medical device, pharmaceutical and biotechnology companies are required to operate under a Quality Management system. For medical devices, ISO 13485 serves as a Quality Management System. Likewise, the FDA Code of Federal Regulations 21 CFR Part 211 for finished pharmaceuticals.

Organisational controls for Data Integrity can address:

- ➢ Assessment of GxP computerised systems
- ➢ Management of GxP computerised systems
- ➢ Electronic Records Implementation and handling
- ➢ Use of Electronic signatures

> ➤ Quality Risk Management

Operational Factors

Operational factors refer to process or manufacturing errors, deviations or non-compliance to established procedures that may impact data integrity.

GxP data handling activities should be designed to limit human intervention. As with human intervention there can errors or omissions. Furthermore, it may call in to question the reliability of the data.

Mistaking proofing methodologies should be developed to avoid human error related breaches in data integrity. As with any system or technology training is a fundamental step. Building upon training, exposure to GxP data systems and On the job training all play a part in delivering a system that is robust and meets regulatory requirements. It is important to remind ourselves that while regulations are the driving force to comply with Data integrity, it is for the protection and safety of the patient or end user of the product, medicine or treatment.

CHAPTER 9

Guide to Inspection Of
Validated Cleaning Processes

Introduction

Cleaning Validation Programs are important requirements for both bulk pharmaceutical processing and Biotechnology. As with validation of other processes, there may be more than one way to validate a cleaning process. Once the manufacturer can establish inspection consistency and repeatable outcomes that ensure pre-determined acceptable criteria are met, a cleaning procedure can be deemed effective. This is a data driven process which should support claims of consistent outcomes.

Not only the FDA have an expectation that cleaning procedures (processes) be validated, PIC/s, ICH, Eudralex and WHO guidance and requirements also specify the need to validate cleaning procedures.

Figure: Components of Cleaning Program

What is Cleaning?

Cleaning can be defined as the process of removing potential contaminants from process equipment and maintaining the condition of equipment so that the equipment can be safely used for subsequent product manufacture. It is complicated by many different chemicals used to produce medicinal drug products and other chemical agents used in the manufacturing process or in the cleaning process.

Why Clean Equipment or Products?

Facilities can be multi-product facilities i.e. the same equipment is shared over different products. However, dedicated facilities also require cleaning evaluation and strategies. Cleaning minimises the transfer (or carry-over) of one product into another product, to an acceptable level, by means of product residue.

Some product residues are considered so toxic/potent if carried-over to another product, that they are required to be manufactured in a dedicated facility e.g. penicillins

Equipment: Clean-in-place often abbreviated to CIP, allows equipment cleaning to occur with minimal dis-assembly of equipment. CIP programs allow different products using similar or different materials to be manufactured on the same equipment.

Products: The supply of products and medical devices that are used by patients or Healthcare Professionals must be clean and free of contamination.

Regulatory Requirements: It is the aim of any manufacturer to provide safe and effective products f or use by patients and end users such as doctors and nurses. Companies are granted licenses to supply markets with products based on regulatory compliance and product safety. Cleaning compliance is a key part to achieving a state of compliance and more importantly, supplying safe products.

Verification and Validation

Verification: Verification means confirmation by examination and provision of objective evidence that specified requirements have been fulfilled[1]

When it comes to Cleaning, if the cleaning procedures have not been fully validated, the effectiveness of the cleaning procedure should be verified at the completion of cleaning. This is "verification".

Validation: Validation means confirmation by examination and provision of objective evidence that the particular requirements for a specific intended use can be consistently fulfilled[1].

Definitions

Clean Hold Time (CHT): The total time the parts or equipment are held clean post-cleaning.

Cleaning Agent: the chemical agent or solution used as an aid in the cleaning process.

Cleaning Process Parameters: the parameters that are critical in the cleaning process. Subsequent cleaning process monitoring may or may not utilize these parameters.

Critical Process Parameter (CPP): A control parameter that has a direct relationship to the quality, safety, effectiveness or performance of the intermediate or final product.

Dirty Hold Time (DHT): The total time the parts (or equipment) are held dirty prior to cleaning.

Maximum Allowable Carry Over (MACO): Amount of allowed product residue carry-over from lot-to-lot, batch-to-batch, etc. This limit is based on the lowest of:

(1) Limited based on Toxicity,
(2) Limit based on Smallest Therapeutic Dose, and

---

[1] 21CFR820. 3

(3) Worst Case Dose Methodology

Residue: Substance left on surfaces of equipment after cleaning that may pose as risk for subsequent use. Example: residues that may require cleaning include: product, excipients, raw materials/intermediates, non-volatile solvent, non-intrinsic cleaning agents such as detergents, etc.

Worst Case conditions: considered to pose the greatest chance of process or product failure. The highest or lowest value of a given control parameter or set of parameters.

Validation: confirmation via documented evidence that the particular requirements for a specific intended use can be consistently fulfilled under anticipated conditions.

Verification: confirmation by examination and provision of objective evidence (i.e. Documentation) that the specified requirements have been fulfilled.

Visual Inspection: With regards to cleaning, visual inspection should be completed by appropriately trained and experienced personal on completion of equipment/process clean down.

Surfaces should be visibly clean and free or visible residue. Hard to clean places should be examined in particular.

cGMP: Current Good Manufacturing Practices

Concurrent Validation: Validation activities occurring as the same time as one another or concurrent to a product launch.

Prospective Validation: This is when validation is done in advance of commercial manufacturing.

Protocol: An approved document that contains the tests and verifications to be conducted during the validation. Validation Protocols include test methods and test conditions, acceptance criteria and parameters required.

FAT: Factory Acceptance Test – Typically classified as an Engineering activity, the purpose of the FAT is to verify the equipment or system meets the requirements of the URS.

Deviation: An event which results in the failure in respect to the acceptance criteria in the protocol.

Process window: The selected operating range of machine setting/parameter that will produce product to meet all quality and product specifications.

Installation Qualification: Establishing through documented evidence that all functionality of the process equipment meet the manufacturer's specification and company requirements.

Equipment Qualification: Providing confidence through documented evidence that the equipment is suitable for the intended use and are capable of consistently operating within set limits and tolerances.

Operational Qualification: Providing confidence through documented evidence that the product can be manufactured to specifications within set limits and tolerances.

(MVP) Master Validation Plan: a governing document which sets out the validation approach and provides details of deliverables. A MVP should be written as soon as possible to do so and should align and reflect with the "current" validation strategy

Precision Cleaning Systems: a precision cleaning system is a piece of equipment that can remove soil or dirt from parts or components. Most Precision Cleaning are made up of several stages, eg Clean-Rinse-Dry. The simplest cleaning system consist of one stage, e.g. an ultrasonic bath, containing heated water.

Clean-In-Place (CIP): is a cleaning method used to clean the inner surfaces of piping, vessels and process equipment without the need for disassembly.

PIC/s: The Pharmaceutical Inspection Convention and Pharmaceutical Inspection Co-operation Scheme (referred to as PIC/S) are two international bodies between countries and pharmaceutical inspection authorities, that co-operative in subjects relating to the field of GMP.

Skid: is essentially a modular process that can be plugged into a process onsite, with little construction or integration. Skids are used as part of Clean-In-Place solutions within Food and Beverage, and Pharmaceutical industries.

Regulatory Requirements

Key regulatory and international publications are included below:

➤ FDA – Food and Drug Administration - Guide to Inspections of Validation of Cleaning Processes

➤ EU GMP – European Commission – Eudralex Volume 4: EU Guidelines to Good Manufacturing Practice, Medicinal Products for Human and Veterinary Use, and Annex 15 (section 10 "Cleaning Validation")

➤ ICH Q7 – International Council on Harmonisation - Good Manufacturing Practice

➤ Guide for Active Pharmaceutical Ingredients (section 12.7 "Cleaning Validation")

➤ ICH Q9 – International Council on Harmonisation - Quality Risk Management

➤ PIC/S PI 006-3 – Pharmaceutical Inspection Co-operation Scheme -Recommendations on Validation Master Plan, Installation and Operational Qualification, non-sterile Process Validation, Cleaning Validation (section 7 "Cleaning Validation")

➤ WHO TRS 937 – World Health Organization - Specifications for Pharmaceutical Preparations; Annex 4: Supplementary guidelines on good manufacturing practices: validation; Appendix 3: Cleaning validation

FDA – Food and Drug Administration - Guide to Inspections of Validation of Cleaning Processes

A Historical Lesson

Historically, the FDA was mostly concerned about the contamination of non-penicillin drug products with penicillins or the cross-contamination of drug products with potent hormones or steroids. One event which increased FDA awareness of the potential for cross contamination due to inadequate procedures was the 1988 recall of a finished drug product, Cholestyramine Resin USP.

In this instance, the bulk pharmaceutical used to produce the product had become contaminated with low levels of both intermediates and degradants.

The cross-contamination in this case was attributed to the reuse of recovered solvents. The recovered solvents had been contaminated because of a lack of control over the reuse of solvent drums. Drums that had been used to store recovered solvents from a pesticide production process were later used to store recovered solvents used for the resin manufacturing process. Some shipments of this pesticide contaminated bulk pharmaceutical were supplied to a second facility at a different location for finishing. This resulted in the contamination of the bags used in that facility's fluid bed dryers with pesticide contamination.

General Requirements

Upon inspection by auditors, the following requirements are expected in order to demonstrate a robust and suitably validated cleaning program (procedure):

➤ Written procedures (SOP's) detailing the cleaning processes used.

➤ Register or list of dedicated equipment. Fluid bed dryer bags are another example of equipment that is difficult to clean and is often dedicated to a specific product.

➤ A written procedure on how cleaning processes are validated

➤ Written validation protocols detailing the sampling procedures, and analytical methods to be used including the sensitivity of those methods and acceptance criteria.

➤ A validation report which is approved in advance of commercial manufacturing. The data generated during the validation should demonstrate that residues have been reduced to an "acceptable level."

Evaluation of Cleaning Validation:

The main focus of an auditor in respect of Cleaning Validation is to evaluate the evidence that aims to demonstrate the effectiveness of the approach and processes used to clean equipment.

The following questions are relevant when evaluating the cleaning process:

> At what point does a piece of equipment /system become clean?
  o This knowledge should be captured in cycle development and development of the cleaning process. Studies may indicate a vessel or piece of equipment requires 3 rinses with hot Water-for-injection at which point it meets acceptance criteria. However, an additional number or rinses may be included to provide a level of confidence in the cleaning process.

> Does it have to be scrubbed by hand?
  o Depending on the drug substances and excipients or other chemicals, residue may tend to physically "stick" to surfaces or behave as tarry or gummy which may require mechanical force to remove them. Or a solvent rinse may be sufficient to remove.

When the cleaning process is used only between batches of the same product a company may only meet the criteria of, "visibly clean" for the equipment. This can often be referred to as a batch-to-batch clean. Such between batch cleaning processes do not require validation. Change over from one product to a different product of different materials requires a more comprehensive clean perhaps with multiple cleans or rinses.

EU GMP – European Commission – Eudralex Volume 4: EU Guidelines to Good Manufacturing Practice, Medicinal Products for Human and Veterinary Use, and Annex 15 (section 10 "Cleaning Validation")

Section 10 of Annex 15 provides a number of bullet points with regard to cleaning validation.

"Cleaning validation should be performed in order to confirm the effectiveness of any cleaning procedure for all product contact equipment. Simulating agents may be used with appropriate scientific justification. Where similar types of equipment are grouped together, a justification of the specific equipment selected for cleaning validation is expected.

A visual check for cleanliness is an important part of the acceptance criteria for cleaning validation. It is not generally acceptable for this criterion alone to be used. Repeated cleaning and retesting until acceptable residue results are obtained is not considered an acceptable approach.

It is recognised that a cleaning validation programme may take some time to complete and validation with verification after each batch may be required for some products, e.g. investigational medicinal products. There should be sufficient data from the verification to support a conclusion that the equipment is clean and available for further use.

Validation should consider the level of automation in the cleaning process. Where an automatic process is used, the specified normal operating range of the utilities and equipment should be validated.

For all cleaning processes an assessment should be performed to determine the variable factors which influence cleaning effectiveness and performance, e.g. operators, the level of detail in procedures such as rinsing times etc. If variable factors have been identified, the worst case situations should be used as the basis for cleaning validation studies.

Limits for the carryover of product residues should be based on a toxicological evaluation. The justification for the selected limits should be documented in a risk assessment which includes all the supporting references. Limits should be established for the removal of any cleaning agents used. Acceptance criteria should consider the potential cumulative effect of multiple items of equipment in the process equipment train.

The risk presented by microbial and endotoxin contamination should be considered during the development of cleaning validation protocols.

The influence of the time between manufacture and cleaning and the time between cleaning and use should be taken into account to define dirty and clean hold times for the cleaning process.
Where campaign manufacture is carried out, the impact on the ease of cleaning at the end of the campaign should be considered and the maximum length of a campaign (in time and/or number of batches) should be the basis for cleaning validation exercises.

Where a worst case product approach is used as a cleaning validation model, a scientific rationale should be provided for the selection of the worst case product and the impact of new products to the site assessed. Criteria for determining the worst case may include solubility, cleanability, toxicity and potency.

Cleaning validation protocols should specify or reference the locations to be sampled, the rationale for the selection of these locations and define the acceptance criteria.

Sampling should be carried out by swabbing and/or rinsing or by other means depending on the production equipment. The sampling materials and method should not influence the result. Recovery should be shown to be possible from all product contact materials sampled in the equipment with all the sampling methods used.

The cleaning procedure should be performed an appropriate number of times based on a risk assessment and meet the acceptance criteria in order to prove that the cleaning method is validated.

Where a cleaning process is ineffective or is not appropriate for some equipment, dedicated equipment or other appropriate measures should be used for each product as indicated in chapters 3 and 5 of EudraLex, Volume 4, Part I.

Where manual cleaning of equipment is performed, it is especially important that the effectiveness of the manual process should be confirmed at a justified frequency."

Ref: EU GMP V4, Annex 15, 2017

ICH Q7 – International Council on Harmonisation - Good Manufacturing Practice:

"Cleaning procedures should normally be validated. In general, cleaning validation should be directed to situations or process steps where contamination or carryover of materials poses the greatest risk to API quality. For example, in early production it may be unnecessary to validate equipment cleaning procedures where residues are removed by subsequent purification steps. (12.70)

Validation of cleaning procedures should reflect actual equipment usage patterns. If various APIs or intermediates are manufactured in the same equipment and the equipment is cleaned by the same process, a representative intermediate or API can be selected for cleaning validation. This selection should be based on the solubility and difficulty of cleaning and the calculation of residue limits based on potency, toxicity, and stability. (12.71)

The cleaning validation protocol should describe the equipment to be cleaned, procedures, materials, acceptable cleaning levels, parameters to be monitored and controlled, and analytical methods. The protocol should also indicate the type of samples to be obtained and how they are collected and labeled. (12.72)

Sampling should include swabbing, rinsing, or alternative methods (e.g., direct extraction), as appropriate, to detect both insoluble and soluble residues. The sampling methods used should be capable of quantitatively measuring levels of residues remaining on the equipment surfaces after cleaning. Swab sampling may be impractical when product contact surfaces are not easily accessible due to equipment design and/or process limitations (e.g., inner surfaces of hoses, transfer pipes, reactor tanks with small ports or handling toxic materials, and small intricate equipment such as micronizers and microfluidizers). (12.73)

Validated analytical methods having sensitivity to detect residues or contaminants should be used. The detection limit for each analytical method should be sufficiently sensitive to detect the established acceptable level of the residue or contaminant. The method's attainable recovery level should be established. Residue limits should be practical, achievable, verifiable, and based on the most deleterious residue. Limits can be established based on the minimum known pharmacological, toxicological, or physiological activity of the API or its most deleterious component. (12.74)

Equipment cleaning/sanitation studies should address microbiological and endotoxin contamination for those processes where there is a need to reduce total microbiological count or endotoxins in the API, or other processes where such contamination could be of concern (e.g., non-sterile APIs used to manufacture sterile products). (12.75)

Cleaning procedures should be monitored at appropriate intervals after validation to ensure that these procedures are effective when used during routine production. Equipment cleanliness can be monitored by analytical testing and visual examination, where feasible. Visual inspection can allow detection of gross contamination concentrated in small areas that could otherwise go undetected by sampling and/or analysis. (12.76)"

PIC/S PI 006-3 – Pharmaceutical Inspection Co-operation Scheme -Recommendations on Validation Master Plan, Installation and Operational Qualification, non-sterile Process Validation, Cleaning Validation (section 7 "Cleaning Validation")

PIC/s provides several pages of recommendation on Cleaning validation. It firstly clearly outlines the principles and purpose of conducting cleaning validation.

"Pharmaceutical products and active pharmaceutical ingredients (APIs) can be contaminated by other pharmaceutical products or APIs, by cleaning agents, by micro-organisms or by other material (e.g. air-borne particles, dust, lubricants, raw materials, intermediates, auxiliaries). In many cases, the same equipment may be used for processing different products. To avoid contamination of the following pharmaceutical product, adequate cleaning procedures are essential.

Cleaning procedures must strictly follow carefully established and validated methods of execution. This applies equally to the manufacture of pharmaceutical products and active pharmaceutical ingredients (APIs). In any case, manufacturing processes have to be designed and carried out in a way that contamination is reduced to an acceptable level.

Cleaning Validation is documented evidence that an approved cleaning procedure will provide equipment which is suitable for processing of pharmaceutical products or active pharmaceutical ingredients (APIs).

Objective of the Cleaning Validation is the confirmation of a reliable cleaning procedure so that the analytical monitoring may be omitted or reduced to a minimum in the routine phase."

Cleaning Validation

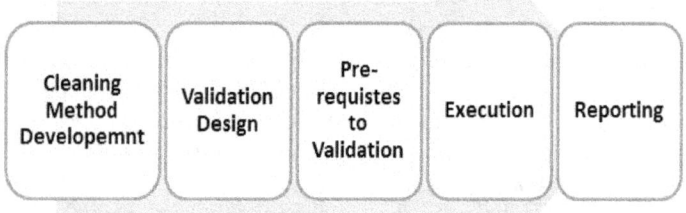

Figure: Lifecycle of Cleaning Validation

The purpose of Cleaning validation is to provide objective evidence that methods and procedures are capable of removing product residues, contaminants, cleaning agents sued, by-products, solvents and degradants to below a pre-determined level.

All contamination is referred to as soiling. A simple method of examining sources of contamination is reviewing the input materials (product ingredients, manufacturing agents etc.) of a process.

Risk assessment must be applied to decide on the extent of cleaning validation studies.

Risk assessment based on product exposure is required to determine the need for validation of facility cleaning procedures (e.g. floors, walls).
Cleaning validation must demonstrate the following with documented evidence (reports and records):

➤ the effectiveness of the cleaning procedure against contamination (chemical or microbiological)

➤ effectiveness of the cleaning procedure against product cross contamination

➤ control of the critical parameters (times, concentration, temperature)

Pre-requisites to Cleaning Validation

Procedures must specify the level of cleaning to be undertaken, cleaning intervals and frequency and the

335

methodology to be utilized. The procedures must be well defined to ensure consistency of operation whether they are manual or automated.

Cleaning validation protocols must describe the test methods to be used (e.g. residue test methods) and limits, microbiological limits and scientific rationales for those limits.

Physico-chemical and microbiological test methods used for cleaning validation must be validated and capable of detecting residues to the required level. Calculations based on scientifically justified limits usually result in impractically high values. For microbial tests the limit of 25 CFU per cm2 (25 CFU/cm2) is typically used for non-sterile manufacture.

As appropriate, development of cleaning procedures must be performed and documented. Worst case considerations must be applied.
Sampling techniques must be defined and authorized. Rationales for using these techniques must be available. An approved sampling plan must be in place.

Execution

Typically, three consecutive applications of the cleaning procedure must be performed and shown to be successful in order to prove a cleaning procedure is validated.

A maximum holding period between end of use and cleaning should be be evaluated as part of the validation study/plan. (Dirty hold time)

A maximum holding period between cleaning and re-use must also be evaluated to determine how long equipment, facilities or systems may remain idle following cleaning. (Clean hold time)

Validation Report

The results of the cleaning validation should be presented in approved cleaning validation report, clearly stating the outcome. Discussion of any issues or non-conformances along with resolutions should also be included.

Clean-In-Place (CIP)

Cleaning Validation for a CIP system design involves the intersection of two similar or different products.
Take a simple example. A Pharmaceutical company manufactures two types of paracetomol caplets (tablets).

Product A contains the Active ingredient paracetomol, preservatives and other excipients.

Product B is also a paracetomol product but it contains an additional ingredient, caffeine. Therefore, product B, is branded differently and marketed with a more"discerning" customer in mind.

Where multiple products are manufactured on the same equipment or machinery, the process is often referred to as non-dedicated.

As with the above example, if the same equipment is used to produce product A and Product B, an intersection of products occurs.

Product A- Cleaning must be effective enough to remove residues to acceptable levels

Product B- when manufacturing commences, the residue levels must not contaminate the product.

Residue is any substance or trace of substance left on equipment or surfaces after cleaning. It is near possible to remove all residue from surfaces, so a residue limit should be medically safe and at a level that does not cause product quality issues or concerns.

"Visibly Clean"

Within any cGMP environment, the requirement to maintain a clean and suitable manufacturing area is key to compliance and ensuring product quality and customer safety. Visual Inspection of the cleaning process must be done before swabbing. Inspection should confirm the equipment is visually clean and dry, and no adverse odours are present.

Upon completion of visual inspection, swanned should then only be taken if required by procedure. For areas that cannot be accessed for visual inspection or swabbing, a rinse sample can be taken in place of a swab.

Sometimes it is not possible to obtain a swab or rinse sample, therefore visual inspection may be the only method used to verify cleaning effectiveness.

In any validation an important theme is to challenge the consistency of a process. Therefore, samples must be representative to ensure a proper picture is painted. Therefore, sampling sites should be taken from "hard to clean" areas as well as "easy to clean" to ensure that samples are representative of the equipment.

Soils and their behaviour

"Soils" are a source of contamination to products and therefore can present a risk to patients or users. Soiled can be introduced by unplanned and unintended events, but they are likely a part of the process, or as a result of a manufacturing agent been used within a manufacturing process, such an example would include coolant, of cutting fluid used in a machining process. The fluid is required to achieve a good surface finish and reduce tool wear. The presence of this soil of parts can potentially be:

- ➢ Dried on during subsequent process step
- ➢ Compacted
- ➢ Dried on during dirty hold time
- ➢ "Baked" on during an oven process

Detergents

Cleaning agents and solvents must be sourced and approve according to a supplier qualification procedure. The following information should be considered for inclusion in a supplier qualification file.

- ➤ Certification to a Quality Management system such as FDA, ISO
- ➤ Supplier questionnaire
- ➤ Product Specification
- ➤ Material Safety Data sheets
- ➤ Change notification policy
- ➤ Expiration dating (format and controls)
- ➤ Onsite audit
- ➤ Statements of suitability
- ➤ Sample Certificate of Analysis

Acceptance criteria for cleaning agent residues should be based on the lowest *LD50* of each chemical in an agent's formulation.

The standard LD50 cleaning agent's classification applies to all agents which contain chemical components whose LD50 is greater than 100 mg/kg. For low LD50 cleaning agents, the classification applies to all agents which contain chemical components whose LD50 is less than or equal to 100 mg/kg.

Validation Strategies

In this section we examine Validation Strategies. There are two main approaches for consideration:

(1) Direct Approach

The direct approach consists of validating the cleaning procedure for all pieces of equipment and for all the products made.

(2) Matrix approach (Aka Grouping approach,
    Family approach, Bracketing)

A family or matrix approach can also be used where similar products can be grouped together with a representative

A matrix can be formed and justified by defining a set of parameters and characteristics so that limited number of parameters or quality attributes are representative of the group. It should also focus on the worst case parameters and quality attributes. All of the information that provides a rationale for implementing a matrix approach should be documented with a Risk assessment created.

The following criteria may be considered to define the worst case product in regards to Cleaning Validation.

➢ PDE
➢ Solubility
➢ Cleanability

With regard to Medical Devices, a particular product size of product configuration may be selected to represent the worst-case product. Therefore, by qualifying the worst case, all other products within the family of products would be considered validated.

With regards to Pharmaceutical products, e.g. Solid Dose manufacturing of pain killers, products of a similar chemistry/content can be grouped together.

Summary

Grouping/matrix approach can be done by:

– By product (soil)

– By equipment

– Worst Case

Advantages of Grouping/Matrix Approaches

– Simplify amount of validation work
– Fewer validation runs

Establishing Residue Limits and Acceptance Criteria

Typical target residues on product within Precision Cleaning systems include:

> ➢ Organic Residuals
>
> ➢ Particulate
>
> ➢ Bioburden
>
> ➢ Endotoxin

Typical target residues for CIP systems include:

> ➢ Drug active

> ➤ Cleaning agent

> ➤ Bioburden

> ➤ Endotoxin

> ➤ Degradation products or by products

How are Acceptance levels defined?

Several considerations need to be accounted for when establishing safe and effective residue levels.

– consider the potential effects of target residue on subsequent products or raw materials
– Pharmacology of residue
– Toxicity of residue
– Stability issues

Per European guidance (Reference Human Drug CGMP Notes, 9:2, 2Q 2001) Equipment does not have to be as clean as the best possible method of residue detection or quantification, as absolute cleanliness is required of feasible. However, it should be as clean as can reasonably be achieved –"to a residue limit that is medically safe and that causes no product quality concerns...."[2]

Historical Context of Limits

[2] Human Drug CGMP Notes, 9:2, 2Q 2001)

Pre- 1993 Industry Acceptance Limits:

- ➤ 1/10th of therapeutic dose
- ➤ 1/50th of Max therapeutic dose
- ➤ Less than smallest therapeutic dose
- ➤ 3ppm (arsenic)
- ➤ 30ppm for cleaning agents
- ➤ Detection limit of method

This approach was Inconsistent from company to company, Arbitrary and not based on Risk.

1993-Eli Lilly Article:

Gary Fourman and Dr Mike Mullin in "Pharmaceutical Technoogy" April 1993 proposed:

- ➤ 1/1000th of a dose in max daily dose
- ➤ 10ppm of product in another product (next product)
- ➤ No residue visible

Uses of the term "limit"

L0 = Daily amount allowed per patient (µg or mg) (L zero)

This has been called the Acceptable Daily Intake (ADI), Acceptable Daily Exposure (ADE), permitted Daily Exposure (PDE). The limit is based safety and toxicity information. Typical values used for L0 include: 0.001 of minimum daily dose of active based on toxicity information.

L=1 Concentration in next product

PIC/s guidance suggests a maximum of 10ppm.

L=2 Absolute amount in manufacturing vessel train

(mg) [MAC – maximum allowable carryover] – L2

This limit uses the absolute amount in manufacturing vessels. It is calculated by multiplying L1, limit, times the batch size of subsequent product to be manufactured.

L3=Amount per surface area (µg/cm^2)

This is calculated dividing L2 by shared surface area of the equipment train (the sum of surfaces)

L4a = Amount per swab (µg)

The Amount per swab depends on the Limit per surface area (L3) – Swabbed area.

Calculate:
L4a = L3 X Swabbed Area

L4b Conc. in swab extraction solution

Concentration in swab extract depends on:

- o Limit per surface area (L3)
- o Swabbed area
- o Amount solvent present for extraction

Calculate:

L3 X Swabbed Area / solvent extraction amt.

L4c = Conc. in "rinse" water ($\mu g/g$)

L4b change based on volume for extraction
If sample is extracted into 10 g of solvent:

$100 \ \mu g \ / \ 10 \ g = 10 \ \mu g/g$

NOTE: L=0, Daily amount allowed is also known as:

Acceptable Daily intake (ADI),

Acceptable Daily exposure (ADE),

Permitted Daily Exposure (PDE)

Safe Daily Intake (SDI)

Values for L0 can be a minimum daily dose of active 0.001,
or a value based on toxicity data (MSDS sheets etc.)

<u>PDA Technical Reprot No. 29</u>

The PDA technical report proposes the following limits:

- ➤ 1/10th -1/100th for topicals
- ➤ 1/100th – 1/1,000th for Oral Products
- ➤ 1/1,000th – 1/10,000th Injections & Opthalmics
- ➤ 1/10,000th – 1/100,000th for Research or Investigational Products

Calculation of MACO (Maximum allowable carryover)

There are two steps in the calculating MACO residue levels. First, it is necessary to **calculate the MACO from one batch to the next batch.**
The second step is to calculate the **allowable "drug" or "residue" of each unique product contact surface, for each piece of equipment.**

This then provides a **calculation based on the overall equipment train** (aka the equipment line)

The **MACO** is based on three calculations, which are:

- • **MACO** based on Toxicity
- • **MACO** based on the Smallest Therapeutic Drug Dose
- • **MACO** based on Smallest Solution Batch Size

$$NOEL = \frac{LD50 \, x \, NHW}{2000}$$

$$ADI = \frac{NOEL}{SF}$$

$$MAC0 = \frac{ADI \, x \, SSBS}{LNDD}$$

NOEL = No Observed Limit Effect
LD50=Lethal dose of drug
NHW=Nominal Human Weight
2000= Is a constant factor for calculating NOEL
ADI=Allowable Daily intake
SF=Safety factor e.g. 1000
SSBS=Smallest Solution Batch Size

MACO based on Smallest Therapeutic Drug Dose (STDD):

$$Product \; Carry \; Over = \frac{STDD}{SF}$$

$$MAC0 = Product \; Carry \; Over \; x \, \frac{SSBS}{LNDD}$$

MACO based on Smallest Solution Batch Size (SSBS):

$$Worst\ Case\ Number\ of\ Doses = \frac{SSBS}{LNDD}$$

$$MACO = LNDDxWorst\ Case\ Number\ of\ \frac{Doses}{SF}$$

Using the above calculations, **the MACO for the equipment train can be determined.**
The MACO for each individual piece of **equipment of surface can then be calculated.**

MACO for each piece of equipment

To calculate the MACO (allowable residue for each piece of equipment) you will need to have the following information available:

-**MACO** (per calculations above)
-**Surface area** of each piece of equipment and the total of the equipment train.

Coverage Testing

Process equipment often contains critical surfaces that need to be cleaned according to validated procedures. Examples include mixing vessels and freeze-dryer chamber. The removal of any residual contamination from these surfaces needs to be demonstrated. This is typically done with an easily detectable tracer such a Riboflavin (e.g. for simple visual detection)

It is important to make the distinction between (1)coverage testing of equipment and (2) coverage testing of components that are subject to cleaning. However, the principle is the same in both instances.

For equipment, coverage testing should be performed as part of equipment qualification for all process-contacting surfaces. Coverage testing verifies that all process-contacting surfaces are wetted by cleaning liquids and to identify any potential blind spots or hard-to-clean locations on the equipment

Locations on equipment that are not adequately cleaned can be identified through riboflavin fluorescence testing using UV light inspection.

Regulatory requirements do not specify a requirement for spray coverage testing. However, as per US Code of Federal Regulations and Eudralex Volume 4 Part II equipment should be of appropriate design to facilitate cleaning.

The Pharmaceutical Inspection Convention and Pharmaceutical Inspection Cooperation Scheme (PIC/S) specify "critical areas (i.e., those hardest to clean) should be identified, particularly in large systems that employ semi-automatic or fully-automatic clean-in-place (CIP) systems."

A 2004 FDA warning letter included two separate mentions of inadequate spray ball coverage: "Your firm failed to establish and follow written procedures to assure the cleaning and maintenance of equipment used in the manufacture, processing, packing, or holding of a drug product 21 CFR 211.67(b) and 600.11(b).

For example, cleaning validation for the clean-in-place (CIP) process vessel which is utilized in the aseptic formulation of trivalent bulk influenza vaccine, did not include an assessment of the spray ball coverage for the vessel. The spray ball is used for cleaning product contact equipment.... The lack of spray coverage testing is mentioned again later in In addition, the cleaning validation did not include an assessment of the spray ball coverage for the tanks"

Ref: FDA Warning Letter issued to Chiron Corporation, December 9, 2004)

Examples of Riboflavin solution strengths

250mg/1 liter of water (1:4 Ratio), generally suitable for spiking components.

2g per 1 liter (1:5),
200mg/L (0.2g/L) solution of riboflavin, suitable for testing interior surfaces of tanks/vessels.

1g of riboflavin in 10 Liters of water 100ppm solution, suitable for coverage testing of equipment surfaces.

Cleaning Validation Protocol

The validation protocol a formal document that is pre-approved prior to its use of execution. It defines the pre-requisites, methods, specific requirements, activities and acceptance criteria for the cleaning validation at hand.
The protocol should address the following:

- ➢ Scope and Objectives
- ➢ Approval by cross functional team
- ➢ Responsibilities
- ➢ Signature and training log
- ➢ Equipment/ Area to be cleaned under study
- ➢ Critical cleaning parameters
- ➢ Sampling methods and sample plan
- ➢ Maximum hold times
- ➢ Acceptance criteria
- ➢ Number of cycles

The dirty hold time (DHT) aka dirty equipment hold time (DEHT) is the time lapse between the end of manufacture and the start of cleaning. The purpose of this time control is to limit the difficulty of the cleaning before residues or remaining product is allowed to dry out.

The clean hold time (CHT) aka clean equipment holding time (CEHT) is the maximum time equipment can sit idle before a re-clean is required prior to its use. The main purpose of this time control is to limit microbial contamination.

The drying time (DT) is another time control that aims to limit microbial growth within vessels or equipment. It is important to dry equipment immediately after it has been cleaned.

PIC/S Guidance on Limits

The Pharmaceutical Inspection Convention and Pharmaceutical Inspection Co-operation Scheme (jointly referred to as PIC/S) are two international instruments between countries and pharmaceutical inspection authorities, which provide together an active and constructive co-operation in the field of GMP.[3]

The most important point to remember when it comes to limits is that residues meet a pre-defined criteria, the most stringent criteria as listed below.

(a) No more than 0.1% of the normal therapeutic dose of any product should appear in the maximum daily dose of the following (next) product,

(b) No more than 10 (parts per million, ppm) of any product will appear in another product, (this value is not always the default)

(c) No quantity of residue should be visible on the equipment after cleaning procedures are completed. Spiking studies should determine the concentration at which most active ingredients are visible. [4]

The method of determining residue limits of active ingredients is based on a approached developed by Fourman and Mullen (1993), and is referenced in PIC/s guidance amongst other publications.

---

[3] http://www.picscheme.org/

[4] http://www.picscheme.org/publication.php?id=4

Test Methods

It is important to consider test methods, and test method validation early on in the validation lifecycle. A Test Method is a process or an action used to verify that a product feature or particular requirement meets a predefined specification. Test methods can be physical or analytical in nature. Test Method validation should be completed in advance of Cleaning as the test method is used to verify the outputs of such cleaning validations.

ICH, Q7, Validation of Analytical Methods

"Analytical methods should be validated unless the method employed is included in the relevant pharmacopoeia or other recognized standard reference. The suitability of all testing methods used should nonetheless be verified under actual conditions of use and documented. (12.80)

Methods should be validated to include consideration of characteristics included within the ICH guidances on validation of analytical methods. The degree of analytical validation performed should reflect the purpose of the analysis and the stage of the API production process. (12.81)

Appropriate qualification of analytical equipment should be considered before initiating validation of analytical methods. (12.82)

Complete records should be maintained of any modification of a validated analytical method. Such records should include the reason for the modification and appropriate data to verify that the modification produces results that are as accurate and reliable as the established method. (12.83)"

(Reference: Q7 Good Manufacturing Practice Guidance for Active Pharmaceutical Ingredients Guidance for Industry September 2016.)

Test method validation must address the following parameters in order to demonstrate suitability to the intended use of the method and ensure it is capable of achieving consistent performance.

Definitions

Specificity: is the ability to assess unequivocally the analyte in the presence of components, which may be expected to present. These can include impurities, degradants etc. (ICH Q2)

Linearity: the ability of an analytical procedure (within a known range) to obtain test results that are directly proportional to the concentration of analyte in the sample (ICH Q2)

Range: The interval between the upper and lower concentrations (amounts) of analyte in the sample (including these concentrations) for which it has been demonstrated that the analytical procedure has a suitable level of precision, accuracy and linearity (ICH Q2)

Accuracy: Expression of closeness of agreement between the value which is accepted either as a conventional true value or an accepted reference value and the value found (ICH Q2)

Robustness: a measure of the capability of an analytical procedure to remain unaffected by small, but deliberate variations in method parameters and which provides an indication of its reliability during normal usage (ICH Q2)

Precision: Expression of the closeness of agreement of an analytical procedure (degree of scatter) between a series of measurements obtained from multiple sampling of the same homogeneous sample under the prescribed conditions. Precision may be considered at three levels: repeatability, intermediate precision and reproducibility. Repeatability expresses the precision under the same operating conditions over a short interval of time. Repeatability is also termed intra-assay precision. Intermediate precision expresses within-laboratories variations: different days, different analysts, different equipment, etc. Reproducibility expresses the precision between laboratories (collaborative studies, usually applied to standardization of methodology) ICH Q2)

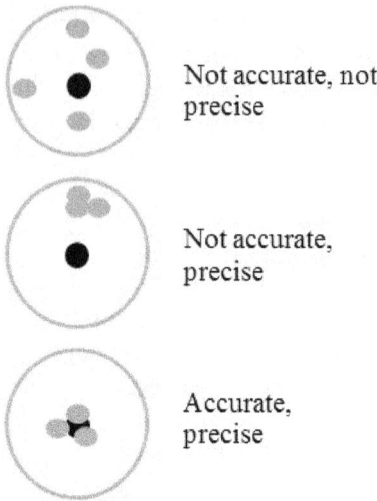

Not accurate, not
precise

Not accurate,
precise

Accurate,
precise

Figure: Illustration of accurate versus precise

Detection limit (LOD): the lowest amount of analyte in a sample that can be detected but not necessarily quantitated as an exact value for an individual analyte procedure (ICH Q2)

Quantitation limit (LOQ): the lowest amount of analyte in a sample which can be quantitatively determined with suitable precision and accuracy for an analytical procedure. The quantitation limit is a parameter of quantitative assays for low levels of compounds in sample matrices and is used particularly for the determination of impurities and degradation products (ICH Q2)

Total organic carbon (TOC) analysis is a fast and effective analytical test method used for cleaning verification and validation in pharmaceutical manufacturing. It is used to test for residues of previously manufactured products (actives and excipients), cleaning detergents, chemicals, solvents, degradants and microbial contaminants. Detergent selection is a critical step in the development a cleaning program. The purpose of the detergent is to remove residues; however, detergent levels remaining post cleaning are undesired. If detergents remain post cleaning they can affect the result of analytical tests.

Cleaning Process Design

For Clean-In-Place (CIP), the Key elements to be considered in design stage include:

— Equipment to be cleaned

— Soils to be removed

— Cleaning methods

— Cleaning agents

— Cleaning mechanisms

— Cleaning parameters

— Residue limits

CIP recipes should include the following information and parameters at a minimum:

➢ type of water (DI, USP Purified, WFI) for pre-rinse,

➤ wash, and post-rinse

➤ volumes, times, flow rates, and pressures

➤ pump speeds

➤ process air blow times

➤ cleaning agent identification, concentration

➤ fill volumes to achieve circuit volume/flow

➤ alarm set points for parameters being monitored

➤ temperature and conductivity (rinses and cleaning solutions)

With regard to Precision Cleaning of Medical Devices, the design inputs are similar to clean in place. However, the focus of the cleaning is on the product that is processed, not the equipment itself.

– Product Material and Product type to be cleaned

– Soils to be removed (e.g. greases, oils)

– Cleaning methods (solvent or aqueous based systems)

– Cleaning agents (detergents, Nitric acid)

– Cleaning mechanisms (ultrasonic, heat, agitation)

– Cleaning parameters (temperatures, times, Ultrasonic frequency)

 – Residue limits (acceptance criteria)

Equipment Considerations

Firstly, for Precision Cleaning Systems, the choice of equipment must be based on the intended purpose of the equipment, for example, will it be used for intermediate cleaning or for final cleaning. As you may expect, there is largely a more stringent acceptance criteria for cleanliness when it comes to final cleaning precision equipment.

For CIP systems, the intended purpose of the equipment is a key design consideration. Materials of construction should be in keeping with the process, maintenance and cleaning regimes associated with it. Material certification should be provided by the supplier or vendor to ensure the correct grade of materials are used from approved suppliers.

In summary, the design should take into account:

➤ Difficult to clean locations
➤ What legacy systems are in place (hence knowledge)?
➤ Materials of construction cleaning agents to be used
➤ cleaning parameters
➤ Individual cleaning or cleaning as an equipment train

Cleaning Agent Approval

(a) Precision Cleaning Equipment

For aqueous based systems, detergents are the preferred cleaning agent used. Detergents are water soluble cleaning agents that "stick" or cause soils and dirt to "bind" together.

Detergents are typically diluted with De-ionised water or suitably clean water. Dosing can range from 5%-10%, though the dose depends on the type of soils present and the equipment. Ultrasonics and temperature and agitation provide for a quicker cleaning cycle.

(b) Clean-In-Place (CIP)

Cleaning agent use with CIP should address the following points:
➢   Effectiveness with regard to API/material
➢   Rinsability
➢   Supplier Certification and Compliance
➢   Stability
➢   Toxicity
➢   Supply and Global availability
➢   EHS compliant

Critical Cleaning Parameters

The key parameters for cleaning can be remembered by using the acronym TACCT.

Time
Action
Cleaning chemistry
Concentration
Temperature

Mixing/flow
Water quality
Rinsing

Other cleaning parameters include flow rate, consideration of turbulence. Water quality and rate of rinsing.

Critical cleaning parameters should be challenged at lower end (least stringent) of the operating window during Validation.

In cleaning development and design studies, parameters can be pushed to the point of failure of below the anticipated operating ranges. By completing more in depth testing in the design phase, process knowledge is gained. It also can allow the cleaning validation to be more focused and avoid

Example: Temperature set-point is expected to be 80 ± 5°C.
– Perform cleaning development studies at set point of 75°C (this covers the expected range and accounts for calibration tolerances)

<u>Cleaning Pipes</u>

The effectively of a cleaning process is influenced by the type of flow within the system, the two types of flow are laminar flow or turbulent flow.

Laminar flow is when fluid particles move in parallel layers, at a constant velocity.

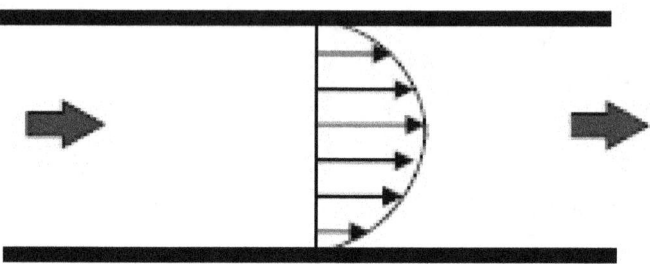

Turbulent flow is when the movement of fluid particles are varying in velocity and direction.

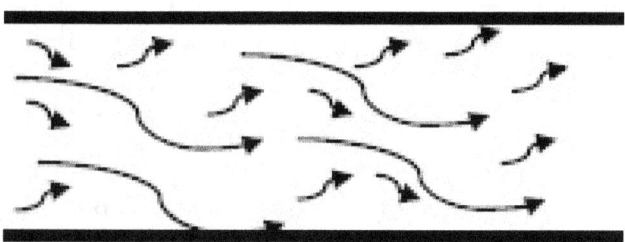

The Reynolds number of a system determines if the flow is turbulent or laminar. A Reynolds Number (Re) greater than 4000 is described as turbulent flow.

$$Re = \frac{316Q}{dK}$$

Q=volumetric flow, (gal/min)

d= internal diameter (inches)

k= viscosity (cP)

Dead Legs

A dead leg is the world of piping terminology refers to an area of piping where there is insufficient flow or a tendency for water build-up or stagnation.

The formal definition of a dead-leg states that Pipelines for the transmission of purified water for manufacturing or final rinse should not have an unused portion greater in length than 6 diameters (6D rule) of the unused portion of pipe measured from the axis of the pipe in use.[5]

---

[5] FDA guidance

Connections and Tie-ins

Precision Cleaning systems and CIP systems both have the necessity to be connected to Utility supplies such as process and de-ionised water. Precision cleanliness will require tie in of water supply and drainage on a continuous or defined frequency. Welding is the preferred permanent method of connecting pipes. Non-permanent connections are also used which allow the disconnection and swap out of piping, vessels and equipment. Orbital welding is a common method of welding when joining piping assemblies and vessels. *(see next section)*

Welding

Stainless steel process piping can be orbitally welded. Quality inspection is typically done real-time by designated quality inspectors using a fibrescope

Welding should meet necessary standards such as the visual weld criteria as detailed in the materials joining part of ASME BPE-2000 Standard. Discoloration of the weld can be evident as a result of the high degree of heat. The discolouration is a result of the oxidation and can result reduce the corrosion resistance of the weld. In general, welds should not exhibit cracks, crevices, or other surface deformities or visual defects. In the event of a weld failure, this can lead to system contamination and would result in the system not been in compliance with 21 CFR 211(a).

Figure: Acceptable weld penetration

Figure: Mismatch or misaligned weld

Figure: Outer Diameter concavity

Figure: Inner diameter concavity, aka suckback

Figure: insufficient penetration

Valves

Clamp-type connections can also be used for non-permanent connections. With regard to the use of Valves. Electromechanical valves that can be PLC controlled are preferred to manual valves. The level of automation, depends on the complexity of the system. For example, a Precision cleanline used to clean metallic hip implants may have 5 or 6 clean and rinse tanks, all fitted with inlets and inlet valves. Having automated control is essential to running a complex line safely and efficiently

Materials of Construction

When it comes to materials of construction, the same selection criteria can be applied to Precision Cleaning Systems and CIP equipment trains. Above all, materials and their surfaces should be non-reactive, non-corrosive and non-porous. Stainless steel of a high grade is often the preferred material of construction. Examples of grades used include 304, 316 and 316L.

For surfaces that are product contacting, material certificates are required to provide evidence that the materials and its constituents are of the correct make-up and suitable grade.

Stainless Steels

Stainless steels (SS) are crystallized solutions of at least 11% of corrosion reducing elements like Chromium and Nickel in Iron. Generally, they are Iron based with 12 to 30% Chromium, 0 to 22% Nickel and minor or no amounts of Carbon, Columbium, Copper, Manganese, Molybdenum, Nitrogen, Phosphorus, Selenium, Silicon, Sulfur, Tantalum, and Titanium.

Casting grades generally are designed with more sulfur to facilitate welding, and have more ferrite (a less corrosion resistant phase) to prevent the formation of micro-cracks on cooling.

Preferred stainless steel for use in the life sciences are manufactured by a VIM – vacuum induction melt followed by a secondary VAR – vacuum arc remelt process with sulfur add back and dispersion in order to minimize inclusions (stringers) and control the amount of sulfur used.

The surface of stainless steel can also be contaminated with the electrolyte solution used in electropolishing if it is used an excessive amount of times or if rinsing steps are not adequate.

This solution builds up Iron and other contaminants that can be transferred to the part being electropolished if the conditions and chemistry are not carefully controlled.

To prevent these problems from occurring all electrolyte solution must be removed from the surface by using a chemically pure water rinse until the conductivity of the rinse from the Stainless steel is equal to the conductivity of the water being supplied for rinsing.

Pressure Testing

Piping or system integration can be required for:

> Precision cleaning systems- where the utilities need to be "tied" into the system

> Installation of Pharmaceutical process within a facility e.g. skid[6] plug in.

After installation, (and before passivation if required) piping systems are pressure tested by filling the system with clean air to 150% of the design pressure or 150psi, whichever is the greatest value. The pressure is then monitored over a 4-hour period to see if there is any drop in pressure.

Passivation

6 see definition in introductory pages

Passivation can be described as the active chemical process used to obtain a uniform Chromium Oxide layer on Stainless Steel (SS) surfaces. The Chromium Oxide layer or film forms a protective coating that gives corrosion resistant properties.

The protective layer naturally forms from the reaction of Oxygen in air with the Chromium on the metal surface but this naturally forming layer can be non-uniform or patchy due to impurities and surface chemistry defects.

When SS is worked such as in welding, machining, mechanical polishing, etc. the uniformity of the naturally forming protective layer can be damaged and oxides of other compounds forming the SS composition can occur. Corrosion can begin at these non-uniform sites and, because SS contains over 60% Iron, the corrosion can proliferate from the surface through the body of the metal if no opportunity for protective layer reformation is given.

There are three passivation processes that are used to enhance the corrosion resistance of Stainless steel.

- ➢ Treatment with oxidizing acids
- ➢ Treatment with chelants
- ➢ Treatment by electropolishing

Layer formation is a dynamic chemical process where the Chromium atoms are combined with Oxygen (and hydroxyl ions in aqueous environments) to form a complex that prevents attack on corrosion prone atoms such as Iron. Nickel and Molybdenum may play a role in formation of the passive film, but the mechanism has not been proven.

Passivation Process

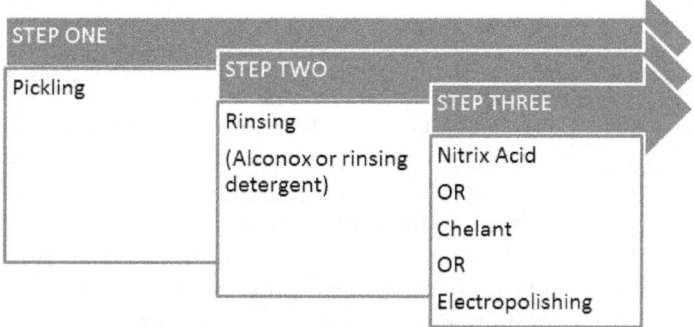

Figure: Passivation- 3 step process.

Passivation must start with a surface free of any oxide scale (including heat tints and oxide corrosion products) and organic contamination (machining lubricants, oils, coolant and grease).

The first step in a successful passivation process is pickling the metal at the mill. Pickling is complete surface element and impurity removal process using a mixture of concentrated of hydrofluoric and nitric acid.

The second passivation step occurs after fabrication and is the removal of organic contamination by washing of the surface with a basic, Trisodium Phosphate (TSP), Alconox or other commerically aviailable, chelant containing, free rinsing detergent at elevated temperatures.

After the organics are removed there are three commonly used processes used to complete passivation. The first uses a hot mineral acid solution (commonly Nitric acid). The second method uses chelants with milder organic acids (Citric acid) and sequestriants. The third is electropolishing.

Mineral Acid:

This is a fast and affordable method of passivation, however its comes with environmental and safety risks.

Nitric Acid is the acid of choice because of its oxidizing properties. The solution is usually heated to facilitate the oxidation reactions. However, the concentration has to be kept below 20% due to the metal surface removal that occurs at higher concentrations. Ph, temperature and conductivity of the acid are also monitored during the process.

Chelant:

Chelant processes are chemical in nature and the materials and their concentrations used can be adjusted to target particular contaminants or all likely residues on the metal.

The published data shows that chelant passivation can achieve Chromium enrichment in the surface of SS down to a depth of approximately 20 angstroms. This is a much higher enrichment and a greater depth of penetration than can be achieved by mineral acid passivation processes.

Chelant processes are proprietary but have the following points in common:

➢ Heating of solutions to assist the chemical kinetics of the processes the passivation solution is usually heated.

➢ Mild organic acid to oxidize the surface Iron to soluble ferric ion and insoluble ferrous ion.

➢ One or more chelants and sequestering agents to capture the ferric and ferrous ions and prevent their precipitation or depositing on the surface of metal.

Electropolishing:

Electropolishing uses a conductive, aqueous salt, reducing acid bath using sulfate and phosphate salts and acids along with a substantial Direct Current power source to remove 0.1 to 2.5 mils of surface metal preferentially from the peaks and high points. It can also remove surface inclusions, free surface Iron and Nickel, carbon and other surface contaminants to a maximum depth of approximately angstroms. This removal will enrich the surface of the SS in Chromium and therefore a highly passive surface is developed.

The article to be electropolished is suspended in the conductive liquid and connected to the anode of the power supply.

A second electrode is also suspended in the conductive liquid and is connected to the cathode. In order to achieve an even metal removal and Chromium enrichment it is important to achieve constant electrical potential across the surface of the article. Electropolishing is limited to improving the surface evenness by approximately 10 Ra.

Stainless Steel and Rouging

Rouging is a form of corrosion found in stainless steels. It can be due to iron contamination of the stainless steel surface due to welding of non-stainless steel for support columns, or other temporary means, which when welded off leaves a low chromium area.

There are three classes of rouging:

> Class I - stainless steel surface and the Cr/Fe ratio of the metal surface beneath such deposits usually remain unaltered.

> Class II - Iron particles originating in-situ on unpassivated or improperly passivated stainless steel surfaces. By their formation the Cr/Fe ratio of the metal surface is altered.

➤ Class III - Iron oxide (or scale) which forms on surfaces in high temperature steam systems. The Cr/Fe ratio of the protective film is usually altered.

(Ref: ASME-BPE)

References

➤ EN 2516:1997 – Passivation of Corrosion Resisting Steels and Decontamination of Nickel Base Alloys

➤ ASTM A380 – Practice for Cleaning, Descaling and Passivating of Stainless Steel Parts, Equipment and Systems

➤ ASTM A967 – Specification for Chemical Passivation Treatments for Stainless Steel Parts

➤ ICH Q3D – International Council for Harmonisation – Guidance for Elemental Impurities

P&ID

Introduction

A piping and instrumentation diagram (P&ID) can be defined as:

1. A diagram which shows the interconnection of process equipment and the instrumentation used to control the process. In the process industry, a standard set of symbols is used to prepare drawings of processes. The instrument symbols are typically based on ISO 10628 International Society of Automation (ISA) Standard (S5.1).
2. The primary schematic drawing used for laying out a process control installation.

They usually contain the following information:

> Process piping, sizes and identification, including:
> Pipe classes or piping line numbers
> Flow directions
> Interconnections references
> Permanent start-up, flush and bypass lines
> tag identifiers),
> Valves and their identifications (e.g. isolation, shutoff, relief and safety valves)
> Control inputs and outputs (sensors and final elements, interlocks)

P&IDs are originally drawn up at the design stage from a combination of process flow sheet data, the mechanical process equipment design, and the instrumentation engineering design. During the design stage, the diagram also provides the basis for the development of system control schemes, allowing for further safety and operational investigations, such as a Hazard and operability study (HAZOP). To do this, it is critical to demonstrate the physical sequence of equipment and systems, as well as how these systems connect.

The most important symbols in relation to cleaning validation include understanding the vessels, tie-ins, valves and instrumentation associated with the equipment under validation.

A valve regulates, directs, or controls the flow of a fluid by opening, closing, or partially obstructing passageways in a piping system. This category includes orifices, and other types of valves. The following pages provide a non-exhaustive list of some examples of common symbols.

Valves:

A valve regulates, directs, or controls the flow of a fluid by opening, closing, or partially obstructing passageways in a piping system. This category includes orifices and other types of valves

Gate
Valve

Gate
Valve

Back
Pressure

Needle
Valve

Above: Gate Valve, Back Pressure Valve and Needle
Valve

Globe
Valve

Control
Valve

Butterfly
Valve

Butterfly
Valve

Above: Globe Valve, Butterfly Valve and Control
Valve

Ball Valve

Diaghragm

**Globe
Valve**

**Check
Valve**

Above: Ball Valve, Diaghragm, Globe Valve and Check
Valve

**Plug
Valve**

**Gate
Valve**

**Check
Valve 2**

**Angle
Valve**

Above: Plug Valve, Gate Valve, Check Valve 2 and
Angle Valve

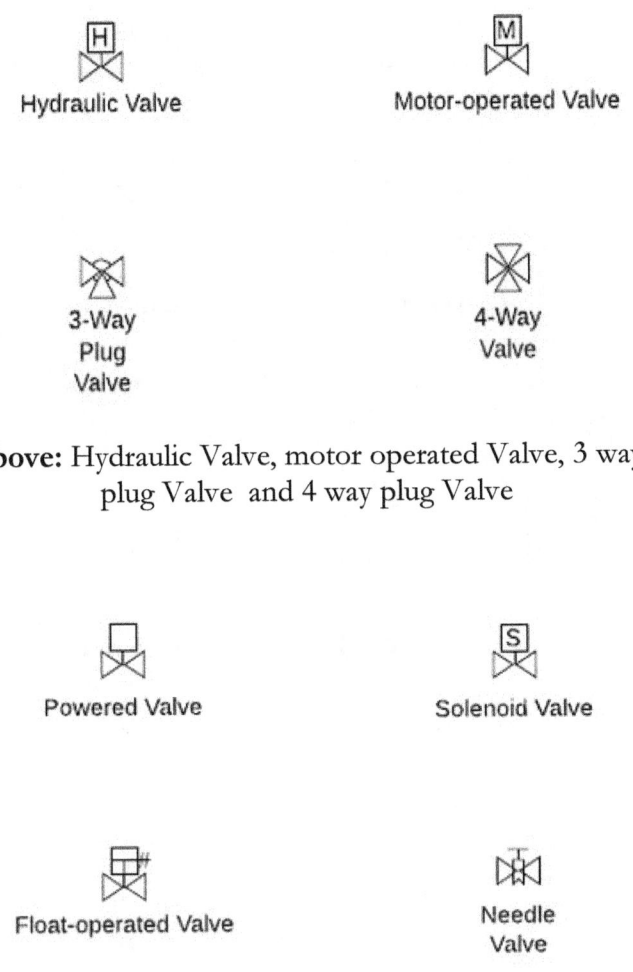

Above: Hydraulic Valve, motor operated Valve, 3 way plug Valve and 4 way plug Valve

Above: Powered Valve, Solenoid Valve, Float-operated valve and needle valve

Vessels

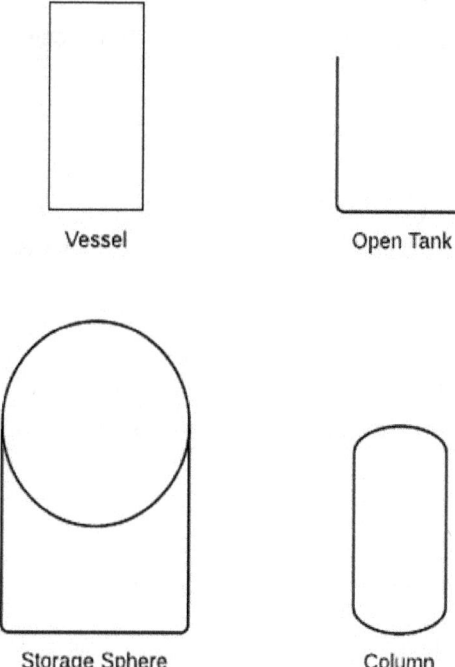

Vessel   Open Tank

Storage Sphere   Column

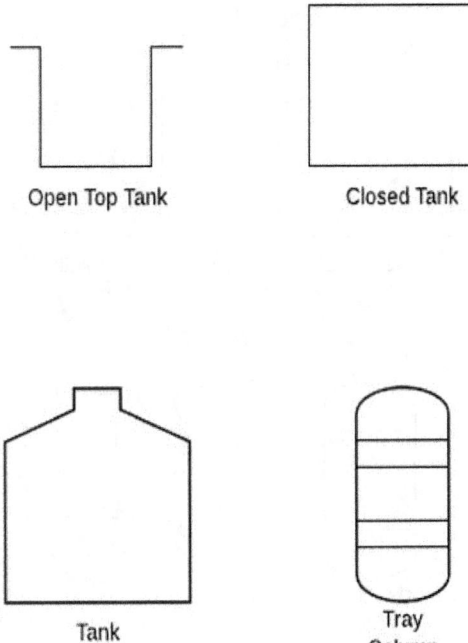

Open Top Tank

Closed Tank

Tank

Tray
Column

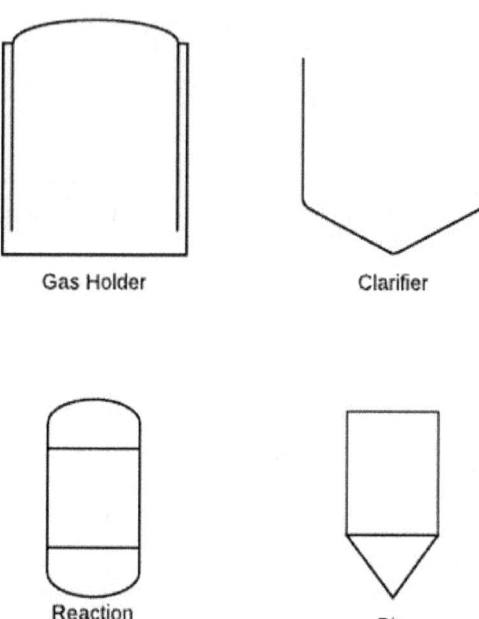

Gas Holder

Clarifier

Reaction
Vessel

Bin

Instruments

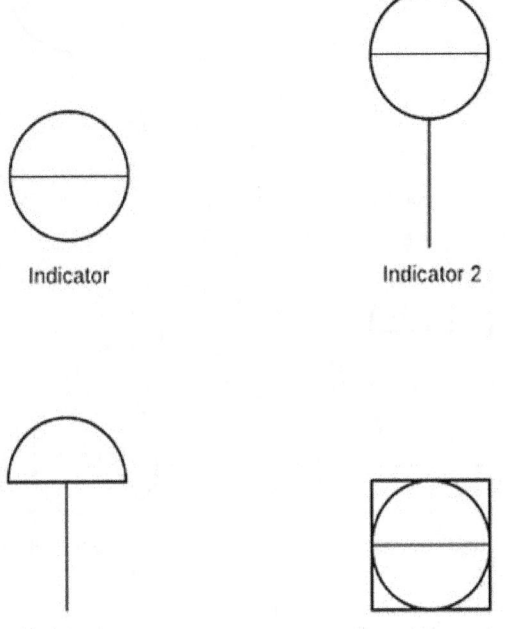

Indicator

Indicator 2

Indicator 5

Shared Indicator

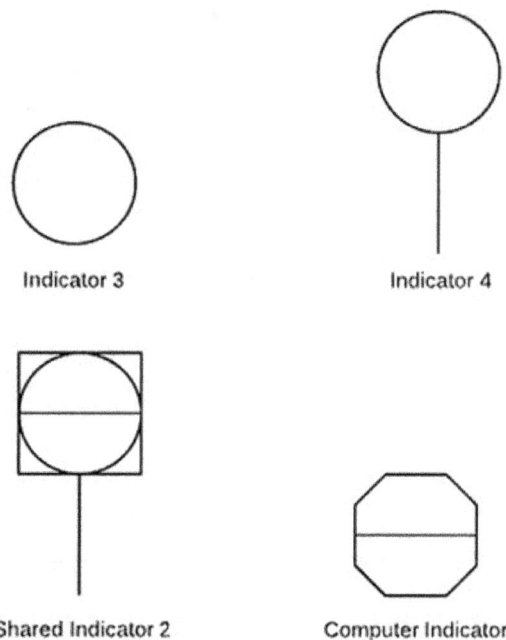

Indicator 3

Indicator 4

Shared Indicator 2

Computer Indicator

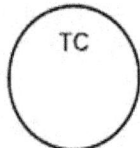

Temp Controller

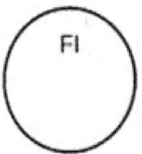

Flow Indicator

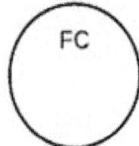

Flow Controller

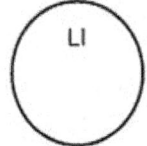

Level Indicator

Sampling

There are two main methods of sampling that are considered to be acceptable, direct surface sampling and indirect sampling (use of rinse solutions).

A combination of the two methods is generally the most desirable, particularly in circumstances where accessibility of equipment parts can mitigate against direct surface sampling

Direct Surface Sampling

(i)   The suitability of the material to be used for sampling and of the sampling medium should be determined. The ability to recover samples accurately may be affected by the choice of sampling material. It is important to ensure that the sampling medium and solvent are satisfactory and can be readily used. Ref: PIC/S PI 006-3

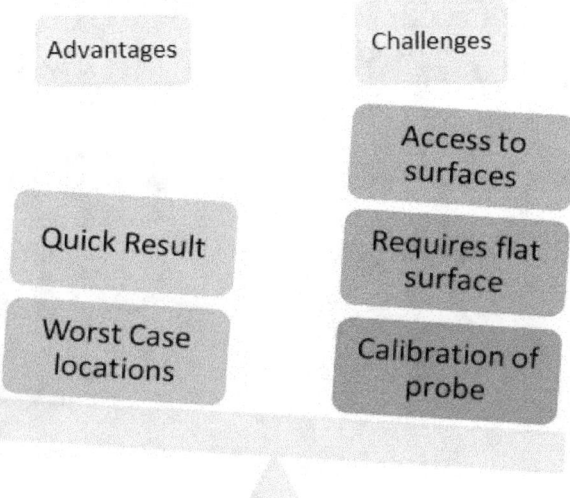

Figure: Direct surface sampling (Pro's and Con's)

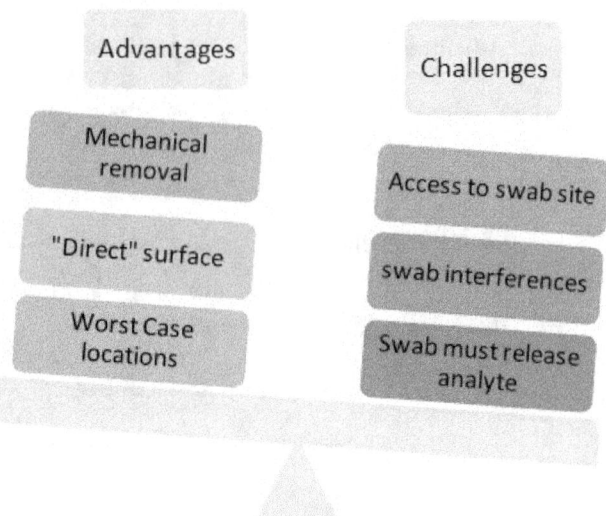

Figure: Swab sampling (Pro's and Con's)

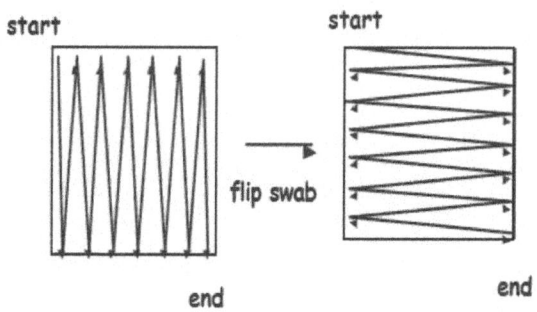

Figure: Method of swab sampling

Rinse Samples

(i)   Rinse samples allow sampling of a large surface
        area. In addition, inaccessible areas of equipment
        that cannot be routinely disassembled can be
        evaluated. However, consideration should be given
        to the solubility of the contaminant.

(ii)   A direct measurement of the product residue or
        contaminant in the relevant solvent should be made
        when rinse samples are used to validate the cleaning
        process. Ref: PIC/S PI 006-3

In rinse sampling a fluid (solvent) is used to rinse and make
contact with all surfaces of the item. The sample is then
tested quantitatively to remove the target residue.

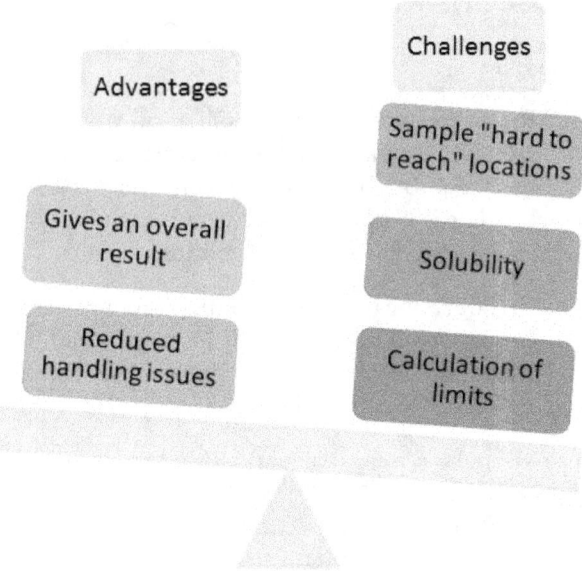

Figure: Rinse sampling (Pro's and Con's)

Sources of Contaminants

Sampling aims to detect any residue drug content or solvents or other soiling left behind after the cleaning process. The visual inspection is also important in identifying any larger contamination of debris.

Microbial sampling is also done to ensure no microorganisms are present in equipment or in areas of production.

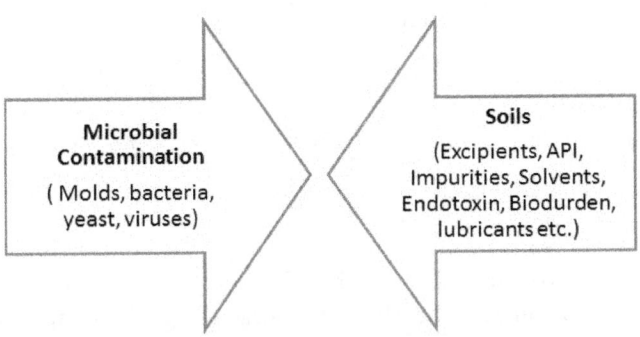

Figure: Sources of Contamination

Microbial Sampling

Microbial sampling of utilities such as Water for injection, Purified Water, Process Air etc. is required to ensure no bacteria, moulds, fungi or yeasts are present which risk patient safety.

Colonies can often be determined by visual inspection based on the attributes and appearance test plates/samples. If visual identification is not possible, the colony should be sent for Gram stain

Visual checks involve assessing plates for:

- colour
- shape
- elevation
- size
- texture
- surface

> edge appearance

Examples

Gram Positive Coccus (Staphylococcus Micrococcus are examples of GPCs)

Spherical bacterium, usually slightly less than 1 μ in diameter, Coccus belongs to the Micrococcaceae family. It is one of th ethree basic forms of bacteria, the other two being bacillus (r odshaped) and spirillum (spiralshaped). A pathogenic coccus can almost always be classified as either a staphylococcus (oc curring in clusters), or a streptococcus (occurring in short orl ong chains). Both staphylococci and streptococci are gram-positive and do not form spores.

The staphylococci are responsible for many serious infection especially Staphylococcus aureus, which is the causativeagen in boils, abscesses, osteomyelitis, and a large variety of other infections. Staphylococci have received muchattention in rec ent years because of the ability of most strains to develop a resistance to antibiotics.

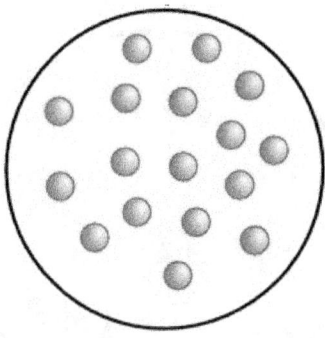

Figure: Single cocci

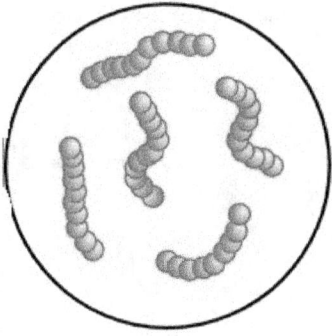

Figure: Streptococci

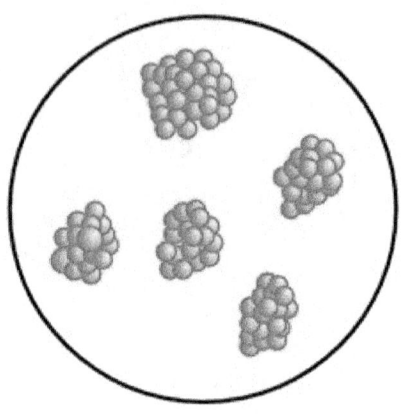

Figure: Staphylococci

Gram positive rod

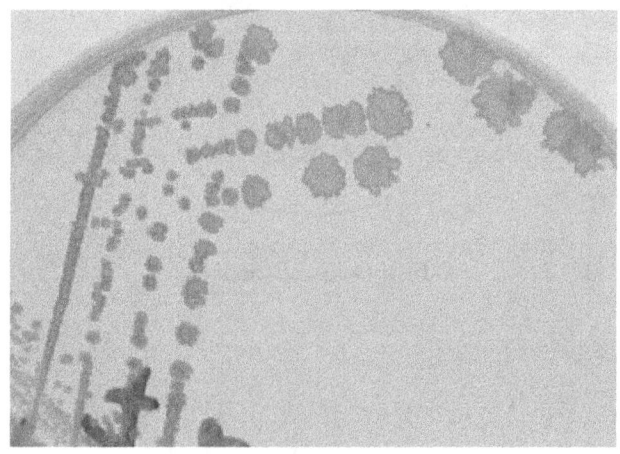

Figure: Gram positive rod

Gram negative rod (E. Coli)

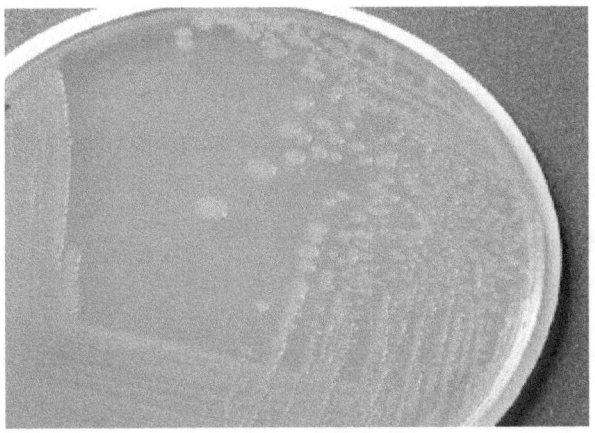

Figure: Gram negative rod (E. Coli)

Gram-negative bacteria are a group of bacteria that do not retain the crystal violet stain used in the Gram staining method of bacterial differentiation.

They are characterized by their cell envelopes, which are composed of a thin peptidoglycan cell wall sandwiched between an inner cytoplasmic cell membrane and a bacterial outer membrane.

Gram-negative bacteria are spread worldwide, in virtually all environments that support life. The gram-negative bacteria include the model organism Escherichia coli, as well as many pathogenic bacteria, such as Pseudomonas aeruginosa and Neisseria gonorrhoeae.

Several classes of antibiotics target gram-negative bacteria specifically, including aminoglycosides and carbapenems.

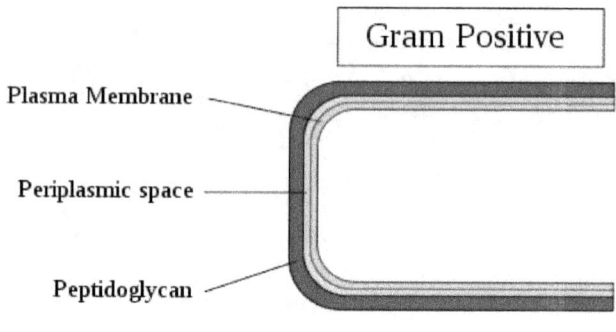

Gram Positive

Plasma Membrane

Periplasmic space

Peptidoglycan

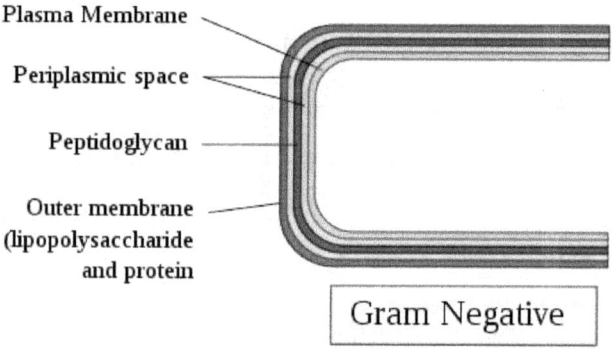

Plasma Membrane

Periplasmic space

Peptidoglycan

Outer membrane
(lipopolysaccharide
and protein

Gram Negative

Figure: Structure of Gram Negative Vs. Gram Positive

Objectionable Microorganism Explained

An objectionable microorganism can be defined as a microorganism that can survive or proliferate in a non-sterile drug product or, when appropriate, intermediates manufacturing steps of sterile and non-sterile processes, and adversely affect the appearance, physicochemical attributes or therapeutic effect of that pharmaceutical product and, A microorganism that, due to its numbers and pathogenicity, can cause infection, allergic response or toxemia in the patient receiving the product.

Factors that may cause a microorganism to be objectionable:

➤ Can it cause infection when present in a medicine depending on the route of administration
➤ Capable to proliferate in a medicine and to exceed the specifications of Total Aerobic Microbial Count (TAMC) or Total combined Yeasts and Moulds Count (TYMC)
➤ Potential Impact the analytical testing
➤ Affect finished product, API, excipients and their respective stability
➤ Produce odors, flavors or undesirable metabolites
➤ Reduction in the efficacy of the medicine

Atypical Microorganisms Explained

An atypical microorganism can be defined as a microorganism that can survive or proliferate in a sterile finished product and can adversely affect the appearance, physicochemical attributes or therapeutic effect of that pharmaceutical product, and,

A microorganism that is not expected to be found in the environment where the product is manufactured and where generally the local flora is dominantly composed of microorganisms of human origin such as gram positive cocci with Staphylococcaceae and Corynebacteriaceae.

Frank Pathogen: Microorganism responsible for infection in healthy individuals (i.e. individuals with normal operative and functional host defense mechanisms) that may bebacquired from exposure to other infected people or animals, environmental reservoirb(exogenous) or the individual's normal (endogenous) microbial flora. [PDA technical report 67]

Microorganism of Concern: A bacterium, yeast or mould that, due to it prominence in product recalls, infection outbreaks, nosocomial infections and the clinical literature, results in a multifactor risk assessment to determine whether the microorganism is objectionable or specified if it is present in a specific non-sterile product or atypical if it is present in a specific sterile product. [PDA TR 67]

Specified Microorganism: Microorganism with limit tests for absence in 1 or 10 g/mL of a finished product, as described in USP<62>/EP 2.6.23 and USP<1111>/EP 5.1.4. [PDA TR 67]

Opportunistic Pathogen: Microorganism responsible for infection in injured, invasively treated or immune-suppressed individuals that typically do not cause infection in healthy individuals, unlike frank pathogens. [PDA TR 67]

Route of Administration: The way in which a drug product or medicinal device is delivered based on the dosage form and therapeutic use. [PDA TR 67]

Utilities

Introduction

The key utilities involved for cleaning include utilities such as water, compressed gases (air, nitrogen etc.) and the heating and cooling of process equipment.
Water quality can impact the effectiveness of pre-rinsing, washing, and final rinsing. Therefore, both the water temperature and quality need to be tightly controlled and monitored. Gases are typically used in order to blowdown or blowout remaining fluids or they are used as a drying step.

The term "Clean Utilities" in the life science industry refers to utilities that have to fulfil regulatory requirements. The most common utility is water, which can be supplied in different pharmaceutical grades of purity. Purified water (PW or PUW), Highly Purified Water (HPW) and Water for Injection (WFI) are the most common. Water quality specifications can be found in the pharmacopaeias, e.g. the US Pharmacopeia. Other clean utilities can also include clean compressed air, clean gasses (e.g. nitrogen, argon and oxygen), and clean steam.

Key Definitions

Alert limit: a value reached when the normal operating range of a critical parameter has been exceeded, indicating that corrective measures may need to be taken to prevent the action limit being reached.

Cleanroom: an area (or room or zone) with defined environmental control of particulate and microbial contamination, constructed and used in such a way as to reduce the introduction, generation and retention of contaminants within the area.

Containment: a process or device to contain product, dust or contaminants in one zone, preventing it from escaping to another zone.

Contamination: the undesired introduction of impurities of a chemical or microbial nature, or of foreign matter, into or onto a starting material or intermediate, during production, sampling, packaging or repackaging, storage or transport.

Point extraction: air extraction to remove dust with the extraction point located as close as possible to the source of the dust.

Pressure cascade: a process whereby air flows from one area, which is maintained at a higher pressure, to another area at a lower pressure.

Relative humidity: the ratio of the actual water vapour pressure of the air to the saturated water vapour pressure of the air at the same temperature expressed as a percentage. More simply put, it is the ratio of the mass of moisture in the air, relative to the mass at 100% moisture saturation, at a given temperature.

Turbulent flow: turbulent flow, or non-unidirectional airflow, is air distribution that is introduced into the controlled space and then mixes with room air by means of induction.

The process of identifying critical utilities can be done with the application of direct impact, in-direct impact and no impact definitions (see previous section *"risk and impact assessment"*). Risk assessments, CQAs and CPPs should also help identify critical utilities. When critical utilities are required as part of manufacturing and processing, the following points should be examined during the requirements and design stage:

- Materials of construction
- Internal surface finishes
- System sizing
- Flow rates, dead legs, drainage etc.

The process of identifying critical utilities can be done with the application of direct impact, in-direct impact and no impact definitions (see previous chapter). Risk assessments, CQAs and CPPs should also help identify critical utilities. When critical utilities are required as part of manufacturing and processing, the following points should be examined during the requirements and design stage:

- ➢ Materials of construction
- ➢ Internal surface finishes
- ➢ System sizing
- ➢ Flow rates, dead legs, drainage etc.

Compressed Air

Compressed air is used for valve actuation, instrument air and process air to name but a few applications. Only the point-of-use filtration and the gas quality instrumentation should be classified as level 1. When flow or pressure is a CPP, the measurement/monitoring should be performed by the system into which the gas is flowing. Additionally, the CQAs and CPPs should be routinely monitored through the calibrated monitoring system. For compressed air, the potential CPPs are listed below. For the physical system being evaluated, the use and the application of the compressed air will determine which (if not all) CPPs are needed to ensure the system produces product of the desired quality.

- ➢ Hydrocarbons
- ➢ Moisture
- ➢ Particulates
- ➢ Temperature

It is important that each point of use has appropriate sterile filters in place. If the filter is not placed directly at the point of use, control and counter measures should be implemented to address any risk of contamination downstream of the filter. Compressed air for bio-pharmaceutical use must be generated using oil free compressors with appropriate temperature controls in place.

Attribute	Clean Compressed Air (impacts product quality)	Sterile Compressed Air (impacts sterile product quality)
Oil content	*Not great than 0.1mg/m³	
Microbiological requirement	Meets requirements of the environmental zone served (e.g. Grade B,C etc)	Sterile
Filtration requirement	Minimum 0.45µm membrane filter	0.2µm membrane filter

*ISO 8573-1 Class 2

Water Systems

Water supply and the associated Water Systems in biotechnology and pharmaceuticals is a vital component of the manufacturing process. It is used to clean equipment and vessels, to cool or heat processing pipes and systems, and in many circumstances certain grades of water is a component of the finished product (e.g. Water-for-injection). Various grades of water service a particular purpose. Some common types include:

➢ Potable water
➢ Soft water
➢ Purified water
➢ Water-for Injection

Water used in process and in cleaning should be pure and free from microbial and chemical impurities. As the water gets easily contaminated by environmental conditions, diligence in the design is essential. Typically water systems are supplied on a continuous loop with recirculation.

CPPs typical for a water system include:
  ➤ Pressure
  ➤ pH
  ➤ Conductivity
  ➤ Level
  ➤ TOC
  ➤ Flow
  ➤ Temperature
  ➤ Resistivity

Water for Injection:

WFI is sterile and pyrogen-free water containing o less than 10 CFU/100ml (Colony Forming Units) with a sample size of between 100 and 300 ml and an endotoxin level < 0.25 EU/ml.

The use of WFI is two-fold. Firstly it can be used for critical processing steps such as washing and rinsing .It can also be used in Injectable products. WFI is a key raw material for sterile intravenous and intradermal products. WFI is produced by Multi Column Distillation Plant (MCDP), and must meet the microbial requirements of regulated bodies.

Clean-in-place (CIP) / Sterilise-in-place (SIP) system

The cleaning of equipment, vessels and process piping is a critical activity. Any residue from a previous production batch needs to be removed in order to avoid cross contamination. CIP and SIP skids often utilised to allow efficient switchover between batches and/or products.

Where possible, manual cleaning should be avoided unless necessary due to the design of a system, or particular location or configuration.

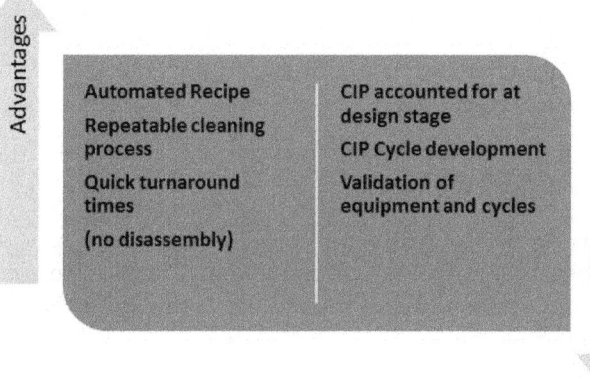

Figure: Advantages and disadvantages of CIP.

Clean Steam

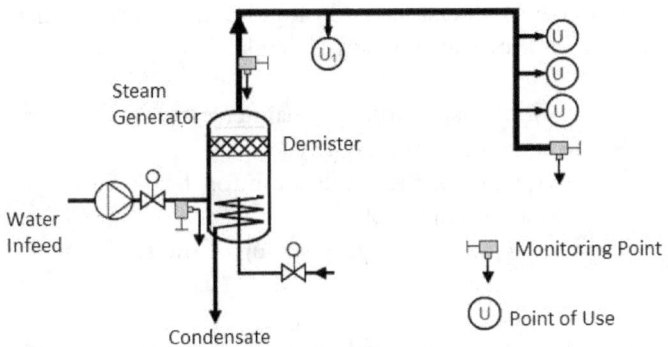

Figure: Simple Clean Steam Generation Piping and Instrumentation

Pure Steam is used in Pharma and Biotech for sterile application, for Autoclave sterilization etc. Distribution piping of Clean Steam is a critical aspect. Improper sizing of pipes may lead impact the production process and loss of time during sterilization.

Clean steam, also referred to as "pure steam", and gases used in manufacturing operations must be of a quality suitable for their intended purpose. The intended use of Clean steam and gases must be understood in order to determine any risks to the patient or product. For example gases end up being part of the product must fulfil the regulatory requirements. Preventative Maintenance and on-going monitoring must be implemented for Clean Steam systems.

> ➤ Routine inspection and maintenance.
> ➤ Frequency of filter change
> ➤ Frequency of the sterilization for the gas distribution system, if applicable
> ➤ Frequency for integrity testing of the sterile filter

Water systems for Purified Water, De-ionised water and Water for Injection (WFI) must provide a consistent and reproducible output. Where there is moisture, there is always a risk of microbial contamination. Therefore, the design of water systems should mitigate against such risks. Good Engineering practices such as using circulation loops, no dead legs and polished surface finishes all work to provide an effective and safe system. The design should also take into account ease of sampling at the point of use. The removal of endotoxins is a requirement for WFI.

On-going sampling monitoring the quality of water is particularly important when water systems are concerned. Procedures should be in place to ensure effective monitoring and testing is maintained. Action limits and acceptance criteria should be clearly documented in approved SOPs or equivalent. Failure to meet limits or acceptance criteria should initiate an investigation. The potential CPPs are listed below for Clean Steam Systems:

> ➤ Conductivity
> ➤ Flow
> ➤ Level
> ➤ Pressure
> ➤ Resistivity
> ➤ Temperature

Design Considerations:

The purpose of a User Requirement Specification (URS) is to define the requirements for the operation and control of the clean steam system.

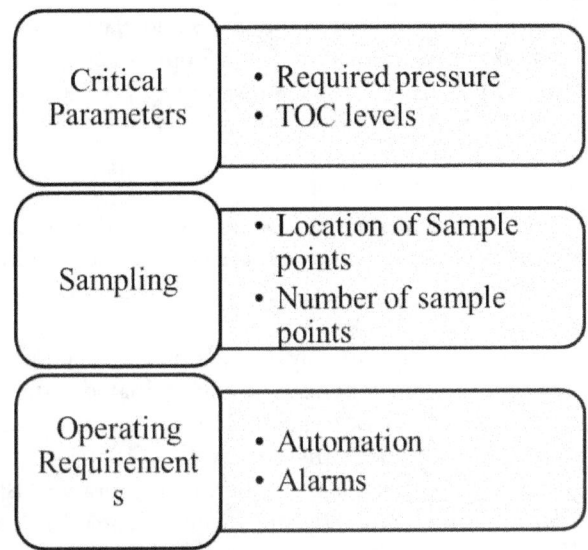

OQ Testing

Operational qualification or OQ is a formal validation activity and as such should be completed per an approved protocol. The purpose of OQ testing is to confirm the operational and functionality of the clean steam system. This should demonstrate that all critical aspects of a U.R.S are fulfilled.

Key verifications include:

➢ Testing of temperatures and operating pressures
➢ Capacity testing (under load)
➢ Steam trap operation
➢ Verification of automated functions and alarms
➢ Check of automation systems, including PCS
➢ Correct function of valves and sampling points

PQ Testing

Due to the high operating temperatures and the associated lethality, clean steam systems are resistant to microbiological contamination.

Issues that arise can normally be attributed to equipment failures with the steam generator or contaminated water been supplied to the system. Bacterial endotoxin testing is used to monitor clean steam systems for both PQ purposes and throughout the lifecycle of the equipment operation. Steam is condensed and sampled and tested. The condensate should meet WFI specifications with the exception of viable total aerobic count.

Clean Steam PQs are commonly completed using a 3 phase approach to testing. The first phase ensures the system consistently operates within the required ranges and the steam provided meets the acceptance criteria. Typically phase one bacterial endotoxin testing and physio-chemical testing is completed over a two week period. For phase 2, the same frequency and type of testing may be applied for an additional 2 weeks. After phase two testing, the system may be available for general use if allowed for within internal company procedures. Phase 2 testing at PQ should also provide a report with all results documented and reviewed.

Phase 3 of PQ is intended to demonstrate the effective and consistent operation of the system over a longer term (approx. 12 months). Sampling is typically performed weekly.

Further reading on Clean Steam

- PIC/S PI009-3 – Pharmaceutical Inspection Co-operation Scheme - Inspection of Utilities

- EN 285 – European Standard - Sterilization, steam sterilizers, large sterilizers

- USP <1231> – United States Pharmacopoeia - <1231> "Water for Pharmaceutical Purposes"

- USP– United States Pharmacopoeia - Monograph "Pure Steam"

EN 285 – European Standard - Sterilization, steam sterilizers, large sterilizers

Case Study 1 –Example of Production Line Switching and Comprehensive Cleaning Procedure/ Requirements

General Safety

1. Switch off the machine prior to commencement of any cleaning operation.

2. LOTO and the authorized worker tag is to be applied by the authorized worker with reference to the applicable LOTO procedure

3. Before starting machine cycles ensure that no operators are working on the machine, that all operators have been informed and acknowledge the machine will be in cycle.

4. Complete GMP permits if required by local and site procedures.

5. Place appropraite signage in-situ

6. Ensure all guards are replaced and checked prior to start up on line.

7. Do not allow water to enter electrical equipment when cleaning.

8. Wear Nitrile gloves when handling Chemical Detergents

150 Tonne V-Blender -Cleaning Procedure

General Cleaning on a daily basis when in production:

1. Wipe down or vacuum the outside surface of the blender in the area around the access covers and around the discharge valve so that no obvious dust accumulation remains

2. Comprehensive Cleaning

A comprehensive clean is carried out:

i)   after 40 calendar days where consecutive operations of the same product take place

ii) between batches of different product

iii) as soon as possible after use if the equipment is to be idle for more than 1 week.

iv) if equipment has been comprehensive cleaned and is idle for more than 7 days then a re-rinse with purified water is sufficient

V-Blender - Manufacturing Areas

Turn off the mains isolator on the Blender control panel.

Note: All equipment must be isolated from the mains before commencing a Comprehensive Clean.

All electrical panels, plugs and sockets must be covered and sealed off. Spraying of water near electrical equipment must be avoided.

Clean the outer and inner surfaces of the Blender and all detachable components as follows:

i) Wash with a solution of made up with the current approved detergent as per cleaning policy and hot water.

ii) A bucket, cloth and mop can be used to apply the detergent solution.

iii) Ensure that all product residue is removed.

vi) Rinse the washed equipment with hot tap water twice using a bucket, cloth and brush. Make sure that there is no visible trace of product on the equipment.

vii) Rinse the equipment with hot tap water twice using a bucket, cloth and mop head.

vi) Rinse the Agitator Bar and internal Blender surfaces with purified water making sure there are no visible traces of detergent on the equipment and allow to air dry.

viii) Wipe off excess water from the outer Blender surfaces.

Re-assemble the equipment only if it is to be used within 7 days of cleaning.

Log the cleaning in the appropriate logbook

A post Comprehensive Clean checklist must also be completed by both the Operator and the Team Leader.

Case Study 2 – Coating Pan

Coating pans are used to coat tablets with thin films of material to enhance a particular property of a tablet. Some coatings are used to slow the dissolution rate, while others may simply act to allow easier swallowing for the patient such as carnauba wax which is a safe, non-toxic and inert ingredient.

The swab sites should be adequate in number and location to provide representative testing, taking into account both easy-to-clean and hard-to-clean locations.

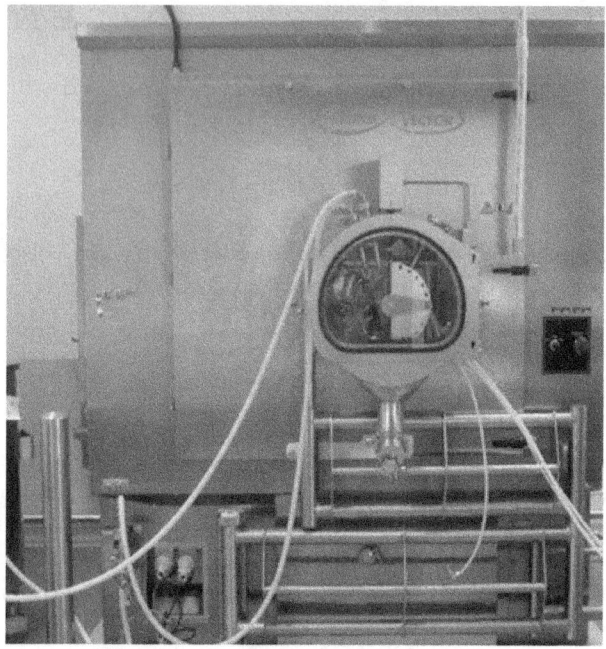

Figure : Coating Pan Exterior View

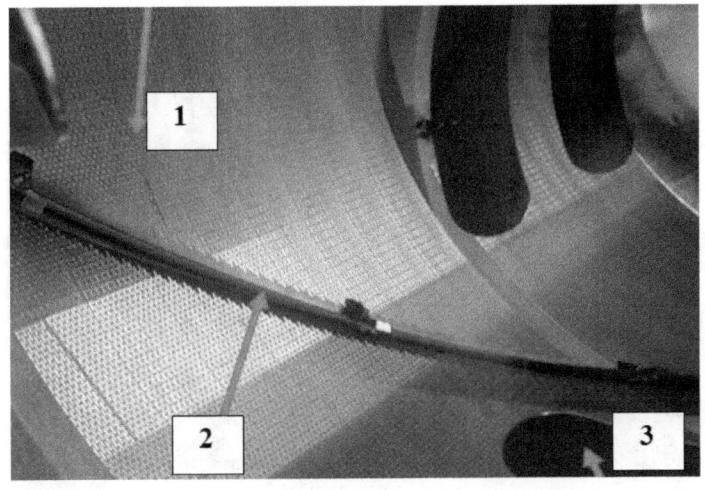

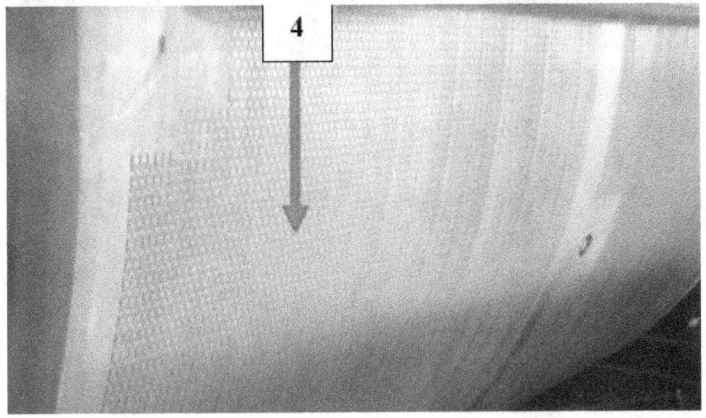

Swab Site 1 – Hard to Clean, interior of perforated pan

Swab Site 2 –Hard to Clean, inner side of wide baffle
Swab Site 3 – Hard to Clean, surface of narrow baffle
Swab Site 4 –hard to Clean, outer surface of pan

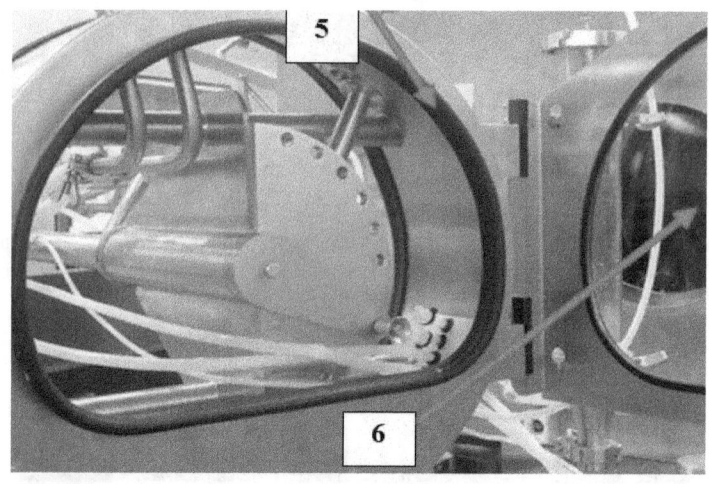

Swab Site 5 – Easy to Clean, rubber seal on door
Swab Site 6 –Easy to Clean, door of pan
Swab Site 7 – Hard to Clean, inside of solution IN line
Swab Site 8 – Hard to Clean, inside ofsolution OUT line

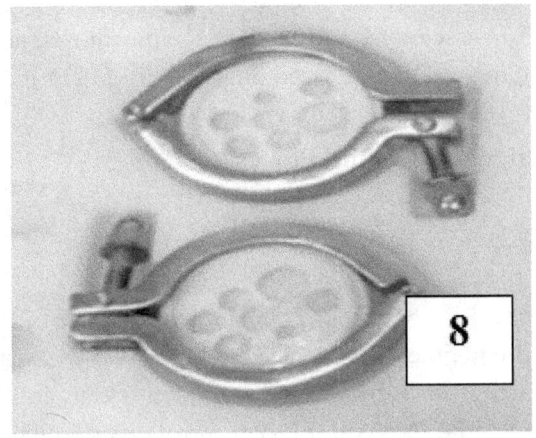

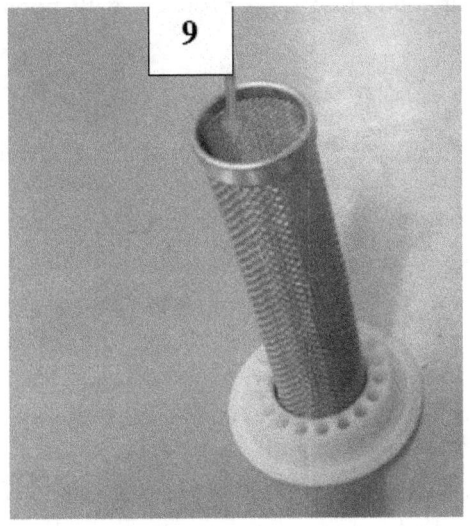

Swab Site 8 – Hard to Clean, rubber seal rim
Swab Site 9 – Hard to Clean, inside the filter housing

Note: Gun assemblies and ancilary equipment may also be included in a validation based on risk.

Inspection of Cleaning Processes

Introduction

As far back as 1963 GMP Regulations (Part 133.4), the FDA required equipment to be clean. The regulations stated that Equipment "shall be maintained in a clean and orderly manner." Nowadays, the FDA is mostly concerned with the cross-contamination of drug products with potent steroids or hormones.

A historical example of cross contamination due to inadequate procedures was the 1988 recall of a finished drug product, Cholestyramine Resin USP.

"The bulk pharmaceutical chemical used to produce the product had become contaminated with low levels of intermediates and degradants from the production of agricultural pesticides. The cross-contamination in that case is believed to have been due to the reuse of recovered solvents. The recovered solvents had been contaminated because of a lack of control over the reuse of solvent drums. Drums that had been used to store recovered solvents from a pesticide production process were later used to store recovered solvents used for the resin manufacturing process.

The firm did not have adequate controls over these solvent drums, did not do adequate testing of drummed solvents, and did not have validated cleaning procedures for the drums.

Some shipments of this pesticide contaminated bulk pharmaceutical were supplied to a second facility at a different location for finishing. This resulted in the contamination of the bags used in that facility's fluid bed dryers with pesticide contamination. This in turn led to cross contamination of lots produced at that site, a site where no pesticides were normally produced.

FDA instituted an import alert in 1992 on a foreign bulk pharmaceutical manufacturer which manufactured potent steroid products as well as non-steroidal products using common equipment. This firm was a multi-use bulk pharmaceutical facility. FDA considered the potential for cross-contamination to be significant and to pose a serious health risk to the public. The firm had only recently started a cleaning validation program at the time of the inspection and it was considered inadequate by FDA. One of the reasons it was considered inadequate was that the firm was only looking for evidence of the absence of the previous compound. The firm had evidence, from TLC tests on the rinse water, of the presence of residues of reaction byproducts and degradants from the previous process."

Ref:https://www.fda.gov/iceci/inspections/inspectionguides/

General Requirements

Written procedures (aka SOP's) is that starting point for application of any cleaning programme within industry. The procedures should give adequate detail on the cleaning processes used for various pieces of equipment.

If different cleaning processes exist for changing between different products then these must be described. Similarly, if there is one process for removing water soluble residues and another process for non-water soluble residues, the written procedure should address both scenarios.

> "FDA expects firms to have written general procedures on how cleaning processes will be validated.

> FDA expects the general validation procedures to address who is responsible for performing and approving the validation study, the acceptance criteria, and when revalidation will be required.

> FDA expects firms to prepare specific written validation protocols in advance for the studies to be performed on each manufacturing system or piece of equipment which should address such issues as sampling procedures, and analytical methods to be used including the sensitivity of those methods.

> FDA expects firms to conduct the validation studies in accordance with the protocols and to document the results of studies.

> *FDA expects a final validation report which is approved by management and which states whether or not the cleaning process is valid. The data should support a conclusion that residues have been reduced to an "acceptable level."*

Evaluating Cleaning Validation

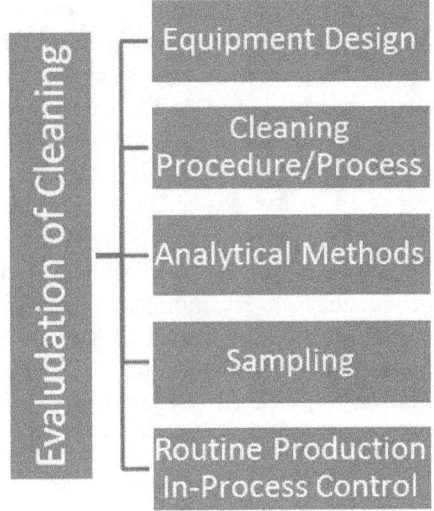

Figure: Key elements of how cleaning validation is evaluated

Equipment Design

It is important to examine the design of equipment, particularly in those large systems that may employ semi-automatic or fully automatic clean-in-place (CIP) systems since they represent significant concern. For example, sanitary type piping without ball valves should be used.

Auditors and inspectors also look at the level of training and experience in cleaning these systems. Also check the written and validated cleaning process to determine if these systems have been properly identified and validated. Always check for the presence of an often critical element in the documentation of the cleaning processes; identifying and controlling the length of time between the end of processing and each cleaning step. This is especially important for topicals, suspensions, and bulk drug operations. In such operations, the drying of residues will directly affect the efficiency of a cleaning process.

After cleaning equipment may be subjected to sterilization or sanitization procedures where such equipment is used for sterile processing, or for nonsterile processing where the products may support microbial growth. Therefore, control of the bioburden through adequate cleaning and storage of equipment is important to ensure that subsequent sterilization or sanitization procedures achieve the necessary assurance of sterility.

Cleaning Process, Procedure and Documentation

Examine the detail and specificity of the procedure for the (cleaning) process being validated, and the amount of documentation required. We have seen general SOPs, while others use a batch record or log sheet system that requires some type of specific documentation for performing each step. Depending upon the complexity of the system and cleaning process and the ability and training of operators, the amount of documentation necessary for executing various cleaning steps or procedures will vary.

When more complex cleaning procedures are required, it is important to document the critical cleaning steps (for example certain bulk drug synthesis processes). In this regard, specific documentation on the equipment itself which includes information about who cleaned it and when is valuable. However, for relatively simple cleaning operations, the mere documentation that the overall cleaning process was performed might be sufficient.

Analytical Methods

The manufacturer should determine the specificity and sensitivity of the analytical method used to detect residuals or contaminants. If levels of contamination or residual are not detected, it does not mean that there is no residual contaminant present after cleaning. It only means that levels of contaminant greater than the sensitivity or detection limit of the analytical method are not present in the sample. The firm should challenge the analytical method in combination with the sampling method(s) used to show that contaminants can be recovered from the equipment surface and at what level, i.e. 50% recovery, 90%, etc. This is necessary before any conclusions can be made based on the sample results. A negative test may also be the result of poor sampling technique.

Sampling

There are two general types of sampling that have been found acceptable. The most desirable is the direct method of sampling the surface of the equipment. Another method is the use of rinse solutions.

Advantages of direct sampling are that areas hardest to clean and which are reasonably accessible can be evaluated, leading to establishing a level of contamination or residue per given surface area. Additionally, residues that are "dried out" or are insoluble can be sampled by physical removal.

Two advantages of using rinse samples are that a larger surface area may be sampled, and inaccessible systems or ones that cannot be routinely disassembled can be sampled and evaluated.

A disadvantage of rinse samples is that the residue or contaminant may not be soluble or may be physically occluded in the equipment. An analogy that can be used is the "dirty pot." In the evaluation of cleaning of a dirty pot, particularly with dried out residue, one does not look at the rinse water to see that it is clean; one looks at the pot.

Routine Production In-Process Control

Monitoring - Indirect testing, such as conductivity testing, may be of some value for routine monitoring once a cleaning process has been validated. This would be particularly true for the bulk drug substance manufacturer where reactors and centrifuges and piping between such large equipment can be sampled only using rinse solution samples. Any indirect test method must have been shown to correlate with the condition of the equipment. During validation, the firm should document that testing the uncleaned equipment gives a not acceptable result for the indirect test.

Establishment Of Limits

The FDA or other regulatory bodies do not intend to set acceptance specifications or methods for determining whether a cleaning process is validated. However, a company's rationale for the residue limits established should be logical based on the manufacturer's knowledge of the materials involved and be practical, achievable, and verifiable. It is important to define the sensitivity of the analytical methods in order to set reasonable limits. Some limits that have been mentioned by industry representatives in the literature or in presentations include analytical detection levels such as 10 PPM, biological activity levels such as 1/1000 of the normal therapeutic dose, and organoleptic levels such as no visible residue.

Useful References

- FDA Process Validation: General Principles and Practices (Appendix A)

- 21 CFR 820.75 Process Validation

- EN ISO 13485:2012 Medical Devices –Quality Management Systems- Requirements for Regulatory purposes (ISO 13485:2003)

- http://www.picscheme.org/publication.php?id=4

- ASME BPE-2000 (Welding) Standard

- FDA – Food and Drug Administration - Guide to Inspections of Validation of Cleaning Processes

- EU GMP – European Commission – Eudralex Volume 4: EU Guidelines to Good Manufacturing Practice, Medicinal Products for Human and Veterinary Use, and Annex 15 (section 10 "Cleaning Validation")

- ICH Q7 – International Council on Harmonisation - Good Manufacturing Practice
- Guide for Active Pharmaceutical Ingredients (section 12.7 "Cleaning Validation")

- ICH Q9 – International Council on Harmonisation - Quality Risk Management

- PIC/S PI 006-3 – Pharmaceutical Inspection Co-operation Scheme -Recommendations on Validation Master Plan, Installation and Operational Qualification, non-sterile Process Validation, Cleaning Validation (section 7 "Cleaning Validation")

- WHO TRS 937 – World Health Organization - Specifications for Pharmaceutical Preparations; Annex 4: Supplementary guidelines on good manufacturing practices: validation; Appendix 3: Cleaning validation

- ICH Q3C(R5) – International Council on Harmonisation – Impurities: Guideline for Residual Solvents

- WHO TRS 986 annex 2 – World Health Organization – WHO Good manufacturing practices for pharmaceutical products: main principles

- WHO TRS 937 – World Health Organization - Specifications for Pharmaceutical Preparations; Annex 4: Supplementary guidelines on good manufacturing practices: validation; Appendix 3: Cleaning validation

- Health Canada Guide-0028– Health Canada - Cleaning Validation Guidelines

- PDA TR29– Parenteral Drug Association – TR29 Points to Consider for Cleaning Validation

Appendix I

Precision Cleaning (Medical Devices)

Load	Clean	Rinse	Clean	Rinse	Rinse	Dryer
	DI Water	DI Water	Nitric Acid with DI Water	DI Water	DI Water	
						Hot Air Blower
	Deterge nt	Heat, Agitatio n & Ultraso nics	Heating and Ultraso nics	Ultraso nics and heating	Ultraso nics and heating	

Precision Cleaning equipment such as Ultrasonic Aqueous cleanlines or solvent based degreasers have one fundamental function in common. This is to remove or reduce particulate, grease and dirt from parts or components. Typically, aqueous or solvent based systems can be used to clean the likes of metallic hip and knee implants, metallic fixation devices, and surgical tools.

Precision Cleaning can be divided into two sub-categories, intermediate cleaning processes and final cleaning processes. As the name may suggest, intermediate cleaning processes are less critical than final cleaning processes and often a reduction in soiling levels (aka organic residuals) is the preferred acceptance criteria. Whereas with final cleaning processes, the level of "cleanliness" is greater and therefore a specific acceptance criteria is applied, based on the application and design of the medical device.

-End-